Rollo's Philosophy [Air]

By

Jacob Abbott

ROLLO PHILOSOPHY

AIR

CHAPTER I
LOST IN THE SNOW

One pleasant morning, very early in the spring, Rollo's cousin Lucy came to call for Rollo to go on an expedition, which they had planned the day before. It was near the end of March, and the snow had become so consolidated by the warm sun in the days, and the hard frosts at night, that it would bear the children to walk upon it. The children called it the crust; but it was not, strictly speaking, a crust, for the snow was compact and solid, not merely upon the top, but nearly throughout the whole mass, down to the ground.

Rollo and Lucy were going to have an expedition upon the crust. Rollo had a sled, and they were going to put upon the sled such things as they should need, and Rollo was to draw it, while Lucy and Nathan, Rollo's little brother, were to walk along by his side.

Rollo's sled was ready at the back door, when Lucy came. Lucy brought with her some provisions for a luncheon, in a basket. This was her part of the preparation. Rollo got his axe, and one or two boards a little longer than the sled, which he said were to make seats. He also had a tinder-box, and some matches, to enable him to make a fire. When all things were ready, the three children set out together.

Rollo drew the sled, with the boards, the basket, and some other things upon it, all bound together securely with a cord. The load appeared to be considerable in bulk, but it was not heavy, and Rollo drew it along very easily. They were not obliged to confine themselves to the roads and paths, for the snow was hard in every direction, and they could go over the fields wherever they pleased. In one place, where the snow was very deep on the side of a hill, they went right over the top of a stone wall.

It was a cloudy day, but calm. This was favorable. The sky being overcast, kept the sun from thawing the snow; but yet their father told them that probably it would begin to grow soft before they came home, and, if so, they would have to come home in a certain sled road, which Jonas had made that winter by hauling wood. He advised them not to encamp at any great distance from the sled road.

They came at last to a pleasant spot on the margin of a wood, near where there was a spring. The rocks around the spring were all covered with snow, and the little stream, which in summer flowed from the spring, was frozen and buried up entirely out of sight. But the spring itself was open, which Rollo said was very fortunate, as they might want some water to drink.

Here they encamped. Rollo cut some stakes, which he drove down into the snow, and contrived to make a rude sort of table with his boards. He also cut a large number of hemlock branches, which Lucy and Nathan dragged out and spread around the table for them to sit upon. Then Rollo built a fire of sticks, which he gathered in the wood. The ground was covered with snow, so that it would have been very difficult for him to have found any sticks, were it not that some kinds of trees, in the woods, have a great many small branches near the bottom, which are dead and dry. These Rollo cut off, and Lucy and Nathan dragged them out, and put them on the fire when he had kindled it. The fire was a little way from the table, with the carpet of hemlock boughs between.

There was a high hill covered with snow at a little distance, and, after they had eaten their luncheon, Rollo said,—

"O Lucy, we will play go up the mountains. There is a hill for us. That shall be Chimborazo."

"Well," said Lucy, "if you will cut us some long staves."

Accordingly Rollo went into the wood, and selected some tall and slender young trees, about an inch in diameter, and cut two for Lucy, two for Nathan, and two for himself. These he trimmed up smoothly, and each of the children took one in each hand. They played that Rollo was the guide, and Lucy was the philosopher. Nathan was the philosopher's servant. Rollo conducted them safely to the summit; but just after they got there, it began to snow.

The snow descended in large flakes, and Rollo was delighted to see it; but Lucy seemed a little anxious. She said that, if there should be much snow, it would make it hard for Nathan to get home, and she thought that they had better go down the mountain immediately, and set out for home. Rollo was rather unwilling to go, but he allowed himself to be persuaded, and so they all came down the mountain together.

They packed up their things as quick as they could, leaving the fire to burn itself out, only Rollo first piled on all the hemlock branches,—which made a great crackling. The snow began to fall faster. The air was full of the large flakes, which floated slowly down, and lodged gently upon the old snow.

The children went along very successfully for some time, but at length Rollo lost his way. The air was so full of snow-flakes, that he could see only a very little way before him; and the old snow covered the ground, so as to hide all the old marks, and to alter the general aspect of the fields so much, that Rollo was completely lost. He, however, did not say anything about it, but wandered on, Lucy and Nathan wondering all the while why they did not get home; until at length they came across a track in the snow.

"O! see this track," said Rollo. "Here is a track, where somebody else has been along with a sled."

"Yes," said Lucy, "some boys, who have gone out to slide, perhaps."

The track was partly obliterated by the snow which had fallen upon it since the boys that made it had gone along. Rollo wondered whose track it could be. He said that he thought it very probable it was Henry's. Lucy thought that it might be the track of some children, that had gone out to find them.

"At any rate," said Rollo, "we will follow the track a little way, and see what it leads to. Perhaps we shall overtake the boys."

Accordingly Rollo turned along in the track, but Lucy stopped.

"No, Rollo," said she, "we must go the other way if we want to find the boys;—the track is going the other way. But never mind," she added, "I don't want to find the boys; I want to go home; so we will go this way."

Rollo went on, secretly pleased to find the track, for he supposed that by retracing it, as he was doing, it would lead him back home. He had, however, a great curiosity to know who could have made it; and in fact the mystery was soon unraveled.

For, after following the track a short distance, they saw before them a large, dark spot upon the snow, and, on drawing near to it, to see what it was, they found it was the place of their own encampment; and the track which they were following was their own track, leading them back to the mouldering remains of their own fire. They had gone round in a great circle, and come back upon their own course. Rollo looked exceedingly blank and confused at this unexpected termination of the clew, which he had hoped was to have led him out of his difficulty. What he was to do now, he did not know.

The fact, however, that they were lost, was no longer to be concealed; and Lucy proposed that they should go into the woods, where the tops of the trees would act as a sort of umbrella, to keep the snow from falling upon them, and wait there until it stopped snowing; and then the air would be clear, and they could find their way out.

"O," said Rollo, "I can easily make a hut of hemlock branches, and we can go into that for shelter."

"But, Rollo," said Nathan, "how do you know but that it will snow all day? We can't stay in the woods all day."

"No," said Rollo; "when it snows in great flakes, it is not going to snow long. Jonas told me so."

So the children went into the woods, and Rollo began to make his hut; but he was soon interrupted, and the attention of all the children was called off by a little bird, which they saw there, hopping about, and appearing benumbed with cold. After some effort, Rollo succeeded in catching her in his cap.

"We'll carry her home," said Nathan; "we'll carry her home, and show her to mother."

"Yes," said Rollo, "I'll carry her in my cap."

"No," said Lucy, "you must have your cap on your head, or you will take cold. Let me carry her in my hands."

"No," said Rollo, "you will have to lead Nathan. I'll tell you what we will do. We will put her into the basket, for a cage."

Lucy, on the whole, liked this plan, and they accordingly put the bird in the basket, and Rollo contrived to make a cover of boughs, to keep her from getting away.

By the time that this was all arranged, the children found, to their great satisfaction, that it had ceased snowing; and they immediately set out for home. Rollo drew the sled, with the basket and bird upon it, secured as before with the cord; and Lucy led Nathan. They now had no difficulty in finding their way, and soon reached home in safety.

They kept the bird until the next day, and then, as it was mild and pleasant, they let her fly away.

QUESTIONS

By what process had the snow become hardened at this time? Did Rollo's father expect that it would continue hard till noon? Why not? Did it continue hard? What prevented its melting? How did Rollo get sticks for his fire? What name did he give to the hill which they ascended? What occasioned the difficulty in the way of Rollo's finding his way home? What conversation took place when he found the track? Was this track in the new snow, or in the old snow? How had it become partly obliterated? How did they carry the bird home?

CHAPTER II
FLYING

Nathan became very much interested in the bird, and that evening, as his father was sitting by the fire, with a book in his hand, which he had been reading, Nathan came up to him, and said,—

"Father, are you busy now,—thinking, or anything?"

"No," said his father.

"Because," said Nathan, "if you are not, I want to read you something out of my little book."

So Nathan's father took him up into his lap, and Nathan opened his little book, and began to read as follows:—

"'With fins for the water, and wings for the air, And feet for the ground, I could go everywhere.'

"Isn't that funny?" said Nathan.

"Rather funny," said his father.

"I wish I had wings," said Nathan.

"Why?" said his father.

"Why, then I could fly."

"That is not certain," said his father. "There are two difficulties which prevent boys from flying. One is, they have no wings; and the other is, they have not strength to use them."

"O father," said Nathan, "I could use them; I am pretty strong. I can wheel Rollo's wheelbarrow."

His father smiled. "Very possibly," said he; "but I do not think that you would be strong enough to use wings, even if you had them."

"Why, at any rate, I am stronger than a bird," said Nathan.

"Yes," said his father, "you have more actual strength than a bird, but not more in proportion to your size. You are absolutely stronger, but not relatively."

"What do you mean by that?" said Nathan.

"Why, you have actually more strength than a bird,—a robin, for instance; you could hold him so that he could not get away; and you could lift more than he could too. But then you are a great deal larger, and you are not as much stronger than he is, as you are larger. If you are a hundred times as heavy as he, you are not a hundred times as strong. That's what I mean by saying that you are absolutely stronger, but not relatively. That is, you are not as many times stronger, as you are larger and heavier. You are absolutely stronger, but not relatively; that is, in proportion to your size and weight.

"Now I can prove to you," continued his father, "that you would not be strong enough to fly with wings, even if you had them. Suppose there was a pole fastened across the room, and another pole just above it; could you pull yourself up, from one pole to the other, by your hands alone, without touching your feet?—Or a ladder," continued his father,—"it will be better to suppose a ladder. Now, if there was a ladder leaning up against a building, could you climb up on the under side by your hands, drawing yourself up, hand over hand, without touching your feet?"

Here Rollo, who was reading in a little chair at the back part of the room, when his father first commenced the conversation with Nathan, but who had been listening for a few minutes past to what his father had been saying, jumped up, and came across the room to his father, and said,—

"Yes, sir, yes, sir; I can. I have done it often in the barn."

"How high up could you go?" said his father.

"O, almost up to the loft," said Rollo. "Only, you see, father, the rounds are too far up. I can't reach up very well. If they were nearer together, I could climb up so, very well."

"Well," said his father, "a bird, when flying, has to climb up in much the same way. He has to pull himself up by the *air*, with his wings, just as you do with your hands and arms, by the rounds of the ladder; only the air is not fixed, like the ladder, but constantly gives way under his wing; and so, to make the case the same, you must suppose that the ladder is not firm, but is floating in the air, and sinks down with your weight, so that you have to climb up faster than you pull the ladder down. Do you think you would have strength enough in your arms to do that?"

Rollo and Nathan looked very much interested in what their father was saying, but they both admitted that they could not climb up such ladders as those.

"The air," added their father, "gives way continually under the bird's wing; and yet they have to pull themselves up by it. And this is very hard. They must either have very large wings, and prodigious strength to use them, so as to pull upon the air with very hard and heavy strokes, or else, if they have small wings, they must have strength to strike very quick and often with them.

"The wings of sparrows move so quick, that you cannot count the strokes; and those of humming-birds, which are smaller still, so fast that you cannot see them. They make a hum."

"I could make my wings go so fast," said Nathan; and he began to imitate the flapping of the wings of a bird, with his arms, as rapidly and forcibly as he could.

"So can I," said Rollo; and he made the same motions. "That is as fast as crows' wings move, when they are flying."

"Yes," said his father, "crows move their *wings* as fast as that, whereas you only move hands and arms. If you had great wings, as long, in proportion, as the crows, you could not move them so fast."

"How large would they be?" said Rollo.

"O, I don't know,—perhaps as big as the top of the dining-table."

"O father," said Rollo, "I don't think they would be as big as that. The crow's wings are not longer than his body, and so mine would not be longer than my body."

"Perhaps you never saw a crow's body," said his father. "His feathers and his tail, which are very light, swell out his body, and make it appear much larger than it really is. I presume his wings, when they are spread, are twice or three times as long as his body. If you had wings in proportion, it would be with the utmost difficulty that you could use them at all. You certainly could not strike the air with them fast enough to pull yourself up by them."

"I did not think that the birds pulled themselves up by the air," said Nathan. "I did not know that the air was anything *real*."

"O yes; it is something real," said his father.

"I've seen birds fly without moving their wings at all," said Rollo.

"Yes," said his father, "and so have I seen a stone."

"A stone!" repeated Rollo.

"What, a stone fly?" said Nathan.

"Yes," replied his father; "did you never see a stone fly through the air, without any wings at all?"

"Why, yes," said Rollo, "when somebody threw it."

"Very well," said his father. "If you set the stone in motion, it will continue in motion for some time, without any wings; and so will a bird."

"But, father, they don't throw birds," said Nathan; and he laughed aloud at such an idea.

"Birds throw themselves," said his father; "that is, they strike their wings upon the air, hard and quick, and thus get into very quick motion, and then they can keep their wings still for a time, and go on, as long as the impulse they have given them lasts. This shows what prodigious strength they have in their wings. They can not only strike the air hard and frequently enough to raise themselves up, and move along, but they can do it so easily, as to get such a velocity, that they can rest their wings for some time, and sail away through the air, only expending the impulse they had accumulated."

Rollo and Nathan were silent. Rollo was thinking how he had seen the swallows sailing swiftly round and round in the air, with their wings spread out motionless by their side.

"So, you see," continued his father, "the difficulty in the way of a boy's flying, is not the want of wings, but the want of strength to use them. It would be very easy to make wings."

"Would it?" said Nathan.

"Yes," said his father. "At least it would not be very difficult. Ingenious mechanics would soon find out modes of making something to answer the purpose of wings, to strike upon the air, if there was the necessary power to work them. The great difficulty in almost all cases in mechanics is, in getting the power; there is very little difficulty in applying it to any purpose it is wanted for. So, you see, next time, Nathan, when you want to fly, you must wish, not that you had wings, but that you were strong enough to use them."

"Well, father," said Rollo, "men are strong enough to paddle themselves along in the water; why can't they in the air?"

"Because," said his father, "water supports them by its buoyancy, and they have nothing to do but to move themselves along upon it. But air cannot support them; and, of course, a great part of the effort which they would make, would be required to keep them up. And then, besides, the water is generally nearly at rest, but the air is generally in a state of rapid motion."

"Why, father," said Rollo, "I'm sure water is sometimes in rapid motion. The rivers run very swiftly, often."

"Yes," replied his father; "but then, when they do, men cannot paddle, or row boats upon them. A current that should run at the rate of four or five miles an hour, would be very hard to row against. But the air is seldom in a state of less motion than that. It is very often moving at the rate of fifteen or twenty miles an hour; sometimes sixty. So, you see, there is a double reason why men cannot fly in the air, as well as paddle on the water."

"If we were only light enough," said Rollo, "to float in the air, then we could fly."

"We could paddle about in it, when it was calm," replied his father, "but that would not be flying."

"Is there anything light enough to float in the air?" said Rollo.

"No," said his father, "I don't think of any visible substance that is."

"What do you mean by visible substance?" said Nathan.

"Why, anything that you can see," replied his father. "There are some other kinds of air, which are lighter than common air, but there is nothing else, so far as I know."

"Why, father, there are clouds. They float, and they are visible, I am sure."

"Yes," said his father. "There is some mystery about the floating of clouds. I don't fully understand it. Clouds are formed of small globules or little balls of water; and water, I should think, whatever the size of the little drops might be, would be heavier than air. And yet they seem to float. If they are large, like rain drops, they fall quickly to the ground. If they are small, like mist, they fall slowly. That I should expect. If they are finer still, like vapor or fog, I should think that they would fall still more slowly; but still I should suppose that they would descend. But they do not appear to descend; they seem to float, nearly at rest; though perhaps all the clouds we see, may be slowly descending all the time, while we do not perceive it."

"The smoke goes up from the top of the chimney," said Rollo.

"Yes," said his father, "there is no difficulty about that. The vapor from a fire is carried up by the warm air, no doubt. Air swells when it is heated, and so becomes lighter, and rises; and the hot air from the top of the chimney carries the vapor up with it, no doubt. After it rises a little way, and becomes cool, it ceases to ascend, but floats away horizontally. Perhaps it begins to descend when it gets cool, though very slowly; and perhaps all clouds are really descending all the time, though too slowly for us to perceive the motion."

"Only," said Rollo, "after a little time, they would get down to the ground."

"Perhaps not," said his father; "for, when they get down nearer the earth, where it is warm, they may be gradually dissolved, and disappear, and thus never reach the earth. I should think they would descend, being composed of globules of water, which, however small, must, I should think, be heavier than air."

"A soap bubble will float in the air," said Rollo.

"I never saw one that would," said his father, "unless it got into a current, which carried it up. A soap bubble—make it ever so thin—shows a tendency to descend, unless you put it out in the open air, where there are currents to carry it up. It descends very slowly, but still it descends. It is heavier than the air. I am not absolutely certain, but I believe there is no visible substance that is lighter than the air; and it is very well for us there is not."

"Why, father?" said Rollo.

"Because, if there were any, they would immediately rise from the earth, and float upwards, till they got up where the air was so light and thin, that they could not go up any higher."

"And so," said Rollo, "we should lose them."

"That would not be all," said his father. "They would float about, above us, and, if there were enough of them, they would form a perpetual cloud over our heads, to keep out the sun, and to make the world dark and gloomy. There seems to have been no way to keep all the solid and visible substances of the earth down upon its surface, but to make them all heavier than the air.

"And thus," continued his father, "all solid substances being heavier than the air, they sink in it, like stones or iron in water. Only those that are very much expanded in surface, sink very slowly, and sometimes almost seem to float."

"What do you mean by *expanded*, father?" said Nathan.

"Spread out," said his father. "An umbrella, for example, when it is spread out, is said to be expanded; other things are expanded in a little different way. A feather is expanded, that is, it is spread out in fine filaments, which extend, in every direction, into the air, all around the stem of it. Things that are expanded take a great deal of air with them when they descend, and so can only descend slowly."

"And water is expanded in a soap bubble," said Rollo.

"Yes," replied his father, "and there is a great deal of air included in it, which all has to be brought down when the bubble itself descends. And thus, you see, the bubble must descend slowly. Water is expanded, too, in clouds; for, in that case, it is divided into millions of small particles, by which it is spread out over a great deal of air, and cannot descend without bringing a large portion of the air with it. Men have contrived, on this principle, to make an apparatus to prevent being hurt by falling from great heights."

"What is it?" said Rollo.

"Why, it is called a parachute. It is a sort of umbrella; in fact, it is an umbrella, only made very large. It is folded up, and fastened under a balloon, just over the car, which the man is in. Then, if the balloon bursts, or any other accident happens to it, and the man begins to fall, the parachute opens and spreads, and then the man falls very slowly. The reason is, that the parachute takes hold of a large mass of air, and brings it down with it; and so it cannot descend very fast."

A few days after this, Nathan said to Rollo, as they were playing in the yard, that he wished that he had a parachute.

"I know where there is one," said Rollo.

"A parachute," said Nathan; "a real parachute?"

"Yes," said Rollo, "or, what is the same thing, a great umbrella."

"Is that just the same?" said Nathan.

"Yes," said Rollo; "for father said that a parachute was in fact only a large umbrella; and father has got a large umbrella in the closet, and I have a great mind to go and get it for a parachute."

"But you haven't got any balloon," said Nathan.

"O, no matter for that," said Rollo.

"Then how are you going to get up into the air?" asked Nathan.

"Why, I can climb up on the shed, and jump off that, and hold the umbrella over my head."

Just at this moment, Rollo's cousin James came into the yard, and Rollo ran to him, to explain to him about the parachute. After describing to him the construction of it, and its use by men who go up in balloons, he said he was going to get his father's umbrella, which would make an excellent parachute.

"And then," continued he, "I am going to get upon some high place, and jump off, and hold the parachute over my head, and then I shall come down as light as a feather."

"O Rollo," said James, "I don't believe you will."

"Yes I shall," said Rollo: "you see the parachute is expanded, and so brings down a great deal of air with it, and this makes it come very slowly. Air is a *real thing*, James, and it keeps the parachute back a great deal."

"James and Nathan both ran towards him, thinking that he must be hurt"—

So Rollo ran off after the umbrella, very much interested in proving to James, by actual experiment, that the air was a real thing. When he came with it, he was himself inclined to make the first experiment from the low side of the shed. He could climb up, by means of a fence at the corner. James advised him, however, to try it first from the end of a woodpile, which was pretty high, but yet not so high as the shed. James was not quite sure that the experiment would succeed, and he was afraid that Rollo might get hurt.

Rollo said that he was not afraid to jump off the shed. He knew the parachute would bear him up. He did not believe but that he could jump off the house with it; and, at any rate, he could jump off the shed, he knew. He accordingly clambered up, and, taking his station upon the eaves, he spread the umbrella over his head, and then jumped off.

Down he came with great violence; his cap flew off in one direction, and his umbrella rolled away in another, as he had to put out both his hands, to save himself, when he reached the ground. As it was, he came down upon all fours, and in such a way, that James and Nathan both ran towards him, thinking that he must be hurt.

"Did you hurt yourself, Rollo?" said James.

"No," said Rollo, "not much."

"I don't think the umbrella did you much good."

"No," said Rollo, as he got up rubbing his elbows, "it didn't, and I don't see what the reason is."

"You came down just as hard as you would without it."

"Yes," said Nathan, "and he almost broke his back; I don't believe the air is any real thing at all."

The fact was, that the umbrella did do *some* good. Rollo did not come down quite so hard as he would have done without it. It retarded his descent a little. But it was not large enough to enable him to descend in safety. When his father said that a parachute was in fact only a large umbrella, he meant a great deal larger than Rollo had supposed. A parachute, such as is used with balloons, is a great deal larger than any umbrella that ever was made.

QUESTIONS

What was Nathan's wish, after he had read his father something out of his book? Did he think that he could fly if he had wings? Did his father think so? What deficiency did his father think was even more important than that of wings? Did Nathan think that a boy was stronger than a bird? Is a boy *absolutely* stronger than a bird? Is he *relatively* stronger? What is the meaning of *relatively* stronger? Would a man be strong enough to work wings that were sufficiently large to bear him up into the air? Would there be any great difficulty in constructing wings for him if he were strong enough?

Is any visible substance lighter than air? What would be the consequence if any of the loose substances about the earth's surface were light enough? What are clouds composed of? What difficulty did Rollo's father point out, in regard to their floating in the air? What is a *parachute*? Describe Rollo's experiment with the umbrella.

CHAPTER III
VALVE MAKING

ONE morning, when Rollo awaked, he heard a sharp clicking against the window.

"Nathan," said he, "Nathan, I believe there is a snow-storm."

But Nathan was too sleepy to hear or understand.

Rollo looked up, but there was a curtain against the window, and he could not see very well. He listened. He heard a low, moaning sound made by the wind, and a continuance of the sharp clicking which he had heard at first.

When he had got up, and dressed himself, he found that there was a violent snow-storm. At first he was glad of it, for he liked snow-storms. But then, pretty soon, he was sorry, for it had been winter a long time, and he was impatient for the spring.

After breakfast, he and Nathan read and studied for two hours, under their mother's direction. When they were released from these duties, Rollo proposed to Nathan that they should go out into the shed, and see how the storm came on. There was a large door in the shed, opening towards the street, where they could stand, protected from the wind, and see the drifts of snow.

They accordingly put on their caps, and went. They found that the snow was pretty deep. It was heaped up upon the fence and against the windows; and there was a curious-shaped drift, with the top curled over in a singular manner, running along from the corner of the shed towards the garden gate.

"Ah," says Rollo, "when it clears up, I mean to go and wade through it."

"And I too," said Nathan.

"O Nathan," said Rollo, "it is over your head."

"Hark!" said Nathan; "who is that pounding in the barn?"

"It is Jonas, I suppose," said Rollo. "I mean to go out and see what he is doing."

"How are you going to get there?" said Nathan.

"O, I can put on my boots," said Rollo, "and go right out through the snow."

"I wish *I* could go," said Nathan.

"Well," said Rollo, "I can carry you on my back."

Nathan clapped his hands at this proposal, being doubly pleased at the prospect of both getting into the barn to see what Jonas was doing, and also of having a ride, on the way.

So Rollo put on his boots, while Nathan went and got Rollo his straps, to fasten his pantaloons around them. When all was ready, Rollo sat down upon the step of the door, in order that Nathan might get on easily.

"We'll play that I am a camel," said Rollo, "and that I'm kneeling down for you to get on."

"Do camels kneel down," said Nathan, "when the men want to ride?"

"Yes," said Rollo; and so saying, he rose laboriously, with his heavy burden upon his shoulders. He staggered along with some difficulty, but yet safely, until he came to the great drift; and, after wallowing into the midst of it, he lost his balance, and both camel and driver rolled over together into the snow. The snow got up under Nathan's sleeve, and he began to cry.

"O Nathan," said Rollo, "don't cry. I'll run and get Jonas to come and carry you in."

So Rollo ran into the barn, and called to Jonas to come quick. Jonas laid down his hammer upon the bench, and followed Rollo. He found Nathan in the snow, and took him up in his arms, and carried him into the barn.

As soon as he got him under cover, he brushed the snow off, and told him not to cry. "I've got a fire in the shop," said he, "and you shall see me do my work. I'm mending the bellows."

So he led Nathan through the barn, and thence along under a shed to a sort of shop-room, where there was a large fireplace and a fire. Rollo put on some sticks, which made a great blaze; and so Nathan soon got warm and dry, and forgot all his troubles. Then Jonas sat him up, upon a high stool, near the bench, where he could see him work. He was just drawing out some of the nails, by which the leather of the bellows was nailed to the sides.

"What is the matter with the bellows?" said Nathan.

"The valve is out of order," replied Jonas.

"The valve," repeated Nathan; "what is the valve?"

"The valve is a kind of clapper," said Jonas. "I will show it to you in a few minutes."

So Jonas proceeded to take off the leather from one of the sides of the bellows. There was a hole in one of the sides, but no hole in the other. Nathan had often noticed the hole, but he did not know what it was for.

"What is the hole for?" said Nathan.

"That is to let the air in," said Jonas.

"What do they want the air to come in for?" said Nathan.

"To make wind of," said Jonas.

"Do they make wind out of air?" said Nathan.

"Yes," said Jonas, "they get the bellows full of air, and then blow it out through the nose, and that makes wind."

"Wind is air, put in motion," said Rollo. "I read it in a book."

By this time, Jonas had taken off the leather so far that Nathan could see into the bellows. He saw that there was a little clapper over the hole, in one of the sides of the bellows.

"Is that the valve?" said he to Jonas.

"Yes," said Jonas.

"What is it for?" said Nathan.

"It is to keep the wind from coming out of that hole."

"Why don't they want the wind to go out of that hole?" said Nathan.

"Because," said Jonas, "they want it to go to the fire,—to blow the fire."

"You see," said Rollo, "it can't go out of the hole, and so it has to go out of the long nose, which is pointed towards the fire."

"What makes it go out at all?" said Nathan.

"Why, when we blow the bellows, we press the two sides together, and that presses the wind out. It can't go out of the hole whence it came in, because the clapper stops it up, and so it goes out the long nose, right into the fire, and makes the fire burn."

By this time, Jonas had got the leather off so far, that he could get at the clapper to mend it. He told the boys that it was old and worn out, and he must make a new one.

"How are you going to make it?" said Rollo.

"You'll see," said Jonas, "if you watch me closely."

So Jonas took some leather, and cut out a piece, of an oblong shape, a little wider than the hole, and about twice as long. Then he laid this down over the hole. It covered it entirely. Then he took some small carpet nails, and nailed one of the ends of the leather down to the board. Then Jonas put his hand down under the board, and run one of his fingers up through the hole, and pushed the leather up a little way.

"There," said he to the boys, "you see I have nailed the leather, so that, when it lies down in its place, it covers the hole completely; and yet I can push it up a little with my fingers, so that there will be an opening."

Then Jonas cut a small leather strap, and nailed one end of it down upon one side of the clapper, and the other end upon the other side of the clapper. He put one little carpet nail into each end of the strap. The strap, when it was nailed, passed directly across the clapper or valve. It was not drawn tight across, but it lay upon the clapper loosely. The ends were nailed tight, but the middle rested loosely upon the clapper.

"Now," said Jonas, "I can push the clapper up a little way, but I can't push it far. The strap keeps it from coming up far."

"But why," said Nathan, "do you want it to go up at all?"

"To let the air in," said Jonas. "When I get the leather all nailed on again, I'll show you the whole operation of it."

"And you can be telling us about it in the mean time," said Rollo.

"Well, then," said Jonas, "when I lift up the upper side of the bellows by the handle, to blow, the air comes in by the hole. The clapper lifts up a little way, and lets it in. Then, when I press down the handle again, it presses the air out through the nose, because it can't go back through the valve hole."

"Why not?" said Nathan.

"Because," said Jonas, "the valve falls down over the hole, and stops it up. It is made so as to lift up easily, and then to fall down and cover the hole exactly, and prevent the air going out the same way it came in. So, as it cannot get out by the valve, it has all to go out through the nose. If the nose were stopped up, it could not get out at all."

"And what then?" said Rollo.

"Why, then," replied Jonas, "you could not bring the two sides of the bellows together again. The air between would keep them apart."

"I should like to try," said Rollo.

"Well," said Jonas; "and there are some other experiments you may perform with it too."

At length, Jonas said that he had got the leather all nailed on, and they might try the experiment. He took hold of the nose of the bellows, and held his thumb near the end of it, ready to stop up the hole.

"Now, Nathan, you may take hold of the handles, and pull them apart as if you were going to blow."

Nathan did so. He pulled the handles apart, and held them open.

"Now," said Jonas, "I will stop up the nose, and the valve will close itself; and then you will find that you cannot bring the sides together again."

So Jonas put his thumb over the hole, and told Nathan to blow.

Nathan pressed hard, and the sides came together again, about as easily as usual.

"What!" exclaimed Jonas with surprise. He did not know what to make of the failure of his experiment.

"There must be a leak somewhere," said he. And he took the bellows out of Nathan's hand to look for it.

He found there was a corner, on the side opposite to the one where he had been working, where the leather was open, he having forgotten to nail it down.

"Ah!" said he, "here is the difficulty. When I have nailed this down, we will try again."

"Is that a leak?" said Nathan.

"Yes," said Jonas. "When you worked the bellows, you pressed the air all out through there. I did not know that that was open. Let me nail this down, and then we will begin our experiment regularly."

QUESTIONS

What was Jonas doing in the shop, when Rollo and Nathan went out to find him? What part of the bellows was out of order? How did he make a new valve? How did he fasten it to its place? Did he nail down only one edge, or both edges? Why did he want the other edge to be left at liberty? How did he prevent its lifting up too far? What was the first experiment which he performed with the bellows, after he had finished the mending? Did it succeed at first? Why not? In working a pair of bellows, where does the air come in? Where does it go out? Why cannot the air escape through the valve where it comes in?

CHAPTER IV
EXPERIMENTS

When Jonas had finished nailing down the corner, he said, "Now there are several experiments, which we can perform with the bellows. I will be the professor, and you two shall be my class in philosophy, and I will direct you how to make the experiments.

"First," said Jonas, "you, Rollo, may take hold of the nose of the bellows with your hand, in such a way as to put your thumb over the end of it, to stop it up, and then let Nathan try to blow."

Rollo did so, and Nathan tried to blow. He found that he could open the bellows very easily; but when he attempted to press the sides together again, he could not. He crowded the handle belonging to the upper side down, as hard as he could, but it would not move.

"What makes it do so?" said Nathan.

"The air inside," said Jonas. "We have stopped up all the places, where it could get out. The valve stops itself. Rollo stops the nose with his thumb, and I have nailed the leather down close, about all the sides. And so the air can't get out, and that keeps you from bringing the sides together again."

Nathan tried again with all his strength. The sides came together very slowly.

"They're coming," said he.

"Yes," said Jonas. "They come a little, just as fast as the air can leak out through the little leaks all around."

"I thought you stopped all the leaks," said Nathan.

"Yes," replied Jonas, "I stopped all the real leaks, but still I can't make it perfectly tight. Some air can escape between the leather and the nails all around, and just as fast as it can get out, so fast you can press the sides together, and no faster."

Here Nathan tried again with all his strength; but he could only bring the sides together very slowly.

"Now comes the second experiment," said Jonas. "While Nathan is trying to press the two handles together, you, Rollo, may run your finger into the hole, and push up the valve a little."

Rollo did so. He pushed up the valve a little with his finger, and that allowed the air to escape through the opening. The consequence was, that the bellows collapsed at once under the pressure which Nathan was exerting upon them.

"There," said Jonas, "you see that when the air is kept in, you cannot bring the sides together; but when I let the air out, then they come together easily."

"Yes," said Nathan; "do it again, Rollo."

So they performed the experiment again. Nathan pulled the handles apart wide, while Rollo kept his thumb over the nose, to keep the air from issuing through. Then Nathan tried to press them together; but he could not, until Rollo put his finger under, and pushed up the valve a little, and then they came together again very easily.

"The air is a real thing, I verily believe," said Nathan.

"Yes," said Rollo, "I know it is. And now for the third experiment, Jonas."

"The third experiment," said Jonas, "is this. Turn the bellows bottom upwards, and try to blow."

Nathan did so. He found that he could work the bellows easily—too easily, in fact; but they did not blow.

"Hold your hand opposite the nose, and see if any wind comes," said Jonas.

They did so; there was no wind, or rather scarcely any.

"The reason is," said Jonas, "that, when the bellows are bottom upwards, the valve hangs down off from the hole all the time, and lets the air all out through the hole in the side; and it can come out more easily there than through the nose, and so it don't blow well."

"Well, Jonas," said Rollo, "that's a pretty good experiment; but what is the next? Let me try the next. Nathan, it is my turn."

"The next experiment, which is the fifth,——"

"No, the fourth," said Nathan.

"The fourth, then," said Jonas, "is to prove what I said to you—that the air, which is blown out at the nose of the bellows, really comes in through the valve. Let me see,—I want something to make a smoke."

"Will not paper do?" said Rollo.

"Yes," said Jonas, "here is some brown paper, which will do." So Jonas rolled it up, and told Rollo to set it on fire, and then, when it was well burning, to step on it with his foot, and put the flame out.

Rollo did so, and the paper lay in a heap, making a great smoke upon the hearth, just before the fire.

"Now," said Jonas, "put the bellows upon its edge, by the side of the paper, so as to have the valve near the smoke, and then hold still a minute, until the smoke comes up steadily by the valve."

When this was done, Jonas told Nathan to take hold of the nose of the bellows, to steady it, so that Rollo could blow. He then directed Rollo to lean the bellows over a little towards the smoke, so that the moving side should not rub upon the hearth, when he began to blow.

"Now," he continued, "if you work the bellows, you will see that the smoke will be drawn in through the valve, and then will come out through the nose."

This experiment succeeded perfectly well, only just in the midst of the interest that they felt, in seeing the smoke come pouring out through the nose, they heard a bell ring at the house. They knew at once that this bell was for Rollo and Nathan; and so the two boys jumped up from the hearth, and ran out to see what was wanted. They went through the shed into the barn, and thence on till they came to the great barn door, where they had come in. There Rollo stopped,—for he did not like to go out into the snow,—and asked Dorothy, who was ringing the bell, what she wanted.

"Where's Nathan?" said Dorothy.

"He's here with me," said Rollo. Nathan was coming along, as fast as he could, through the barn.

"Do you want us?" said Rollo.

"No," said Dorothy, "only we did not know where you were. You may stay half an hour more, and then it will be nearly dinner time."

Dorothy then went in, leaving the boys at the great barn door. The door opened in such a direction, that the wind did not blow in; and Rollo and Nathan looked out for some time, watching the falling snow, and listening to the wind, as it roared through the tops of the trees. At last, when they began to think of returning to the shop, Rollo said,—

"O Nathan, let us go and hide, and then Jonas will not know where we are."

"Well," said Nathan, "we will."

The boys accordingly began to look about the barn for a place to hide. It was a large barn, with stalls for oxen and cows, and cribs for horses, and one or two calf-pens. Then there was a granary in one corner, and a tool-room near it, and lofts and scaffolds above. The boys found plenty of places to hide in, and it took them some time to decide which to choose. At last, they found a good warm place, by some bundles of wheat straw, up in the barn chamber; and they amused themselves by choosing out large straws, and making tubes of them to blow through. They called them their *bellows*.

They entirely forgot that they were hid from Jonas, for nearly half an hour; and then Rollo proposed that they should creep softly down, and see what Jonas was about. So they went down stairs on tiptoe; Rollo first, and Nathan following. They crept softly along to the door leading out into the shed, through which they had to pass in order to get to the shop; and Rollo was going to open this softly, when, to his surprise, he found it fastened.

"Why, Nathan," said he, "this door is fastened."

"How came it fastened?" said Nathan.

"I don't know," said Rollo, "unless Jonas fastened it. I think he must have finished his work, and gone into the house; and so he has fastened this door."

"And now he won't come and find us," said Nathan.

"No," said Rollo, "and we must go out the front door. And I don't care much," he continued, "for it is pretty near dinner time."

The boys then went back to the front door of the barn, and, to their surprise and alarm, found that fastened too.

"What shall we do?" said Rollo; "Jonas has fastened us in." As Rollo said this, his face assumed an expression of great solicitude, and Nathan began to cry.

"Don't cry, Nathan," said he; "we can find some way to get out. But I don't see, I confess, what made Jonas lock us in."

The truth was, that Jonas did not know that the boys were in the barn when he fastened it up. As they did not come back after they had gone to answer the bell, he supposed that they had gone into the house; and when he was ready to come in himself, he shut and fastened the back doors of the barn, as he usually did when he left the shop. He then came around to the front barn door, and although that was on the sheltered side, so that the wind did not blow in, he thought it possible that the wind might change, and so drive the snow in upon the barn floor; and therefore, to make all safe, he thought that he would shut them, too. He accordingly shut the great doors, and put the fid into the staple. The fid is a wooden pin, to be passed through the staple when the doors are shut, to fasten them. The doors cannot be opened again until the fid is taken out.

Rollo went all around the barn, trying to find some place where he could get out; but he could not find any place at all.

"Let us go up stairs," said he, at length, to Nathan.

"O, it will not do any good to go up stairs," said Nathan. "It would kill us to jump out the window."

"I know we can't jump out the window," said Rollo, "but perhaps we can find out some way to get down. O, there is a ladder; I remember now, Nathan, there is a ladder. We can get down from the window by the ladder."

"I shall be afraid to go down the ladder," said Nathan.

"O no," said Rollo, "I will go first, and see if it is safe."

By this time they had reached the barn chamber. There was a window in it, with glass, over the great barn door; but Rollo could not get it open. He told Nathan that, if he could only get it open, and could find a long pole, he could reach it down, and knock the fid out, and so open the great doors. But, with all his efforts, he could not raise the window.

There was another window, which had no glass, but was closed by a wooden shutter, which opened upon hinges like a door. Rollo said he meant to open this window. Now, it happened that this window was upon that side of the barn which was exposed to the wind and storm; and, the moment that Rollo had pushed open the shutter a little way, the wind forced it instantly from his hand, and slammed it back against the side of the barn, with great violence. It almost pulled Rollo himself out of the window.

Nathan looked frightened. Rollo himself looked somewhat astonished at such an unexpected effect; but presently said,—

"Well, Nathan, I rather think that, if you had had hold of that shutter, you would have thought that air was a real thing."

"O, that was the *wind*, Rollo; that was the *wind*," said Nathan.

Rollo did not answer, but went to the ladder, which was standing up against the hay-loft. It was a pretty long, but yet a very light ladder; and Rollo and Nathan succeeded, after some difficulty, in getting it down, and in running the end out of the window. When the lower end reached the ground, the upper end was two or three feet above the bottom of the window; so that Rollo could easily get upon it to descend. The wind and storm, which raged with great violence, were somewhat terrifying; but he knew that the ladder was secure, the upper part being confined in the window; and so he resolutely descended. When he had fairly reached the ground, he looked up, with an expression of great satisfaction upon his countenance, and said,—

"There! now, Nathan, for your turn."

But Nathan was afraid to venture; and Rollo himself was half afraid to have him make the attempt. While they were standing in this perplexity, Rollo heard a voice behind him, calling out,—

"Rollo."

Rollo turned, and saw Dorothy standing by the door.

"What are you doing, Rollo?" said Dorothy.

"I am trying to get Nathan out of the barn," said Rollo.

"How came he in the barn?" said Dorothy.

"Why, Jonas locked us in, and I had to come down the ladder; but Nathan is afraid, and I can't get him out."

"Why don't you go to the door, and let him right out?"

"O," said Rollo, laughing, "I never thought of that. Go down, Nathan," he continued, "to the door, and I will go round and knock out the fid."

So Nathan went down, and Rollo, meeting him there, knocked out the fid, and released him from his imprisonment.

QUESTIONS

What was the first experiment with the bellows, described in this chapter? Why could not Nathan press the two sides of the bellows together, while the nose was stopped? What was the second experiment? What was the effect produced by turning the bellows bottom upwards, as in the third experiment? What was the fourth experiment? What was the use of the smoke of the paper? How were the experiments interrupted? What evidence did Rollo and Nathan have that the air was a real substance, when in the barn chamber?

CHAPTER V
PRESSURE

One evening, just after tea, Rollo came to his father, who was sitting by the side of the fire, and said,—

"Father, I wish we could see the air, as we can the water, and then perhaps we could try experiments with it."

"O, we can try experiments with the air as it is," said his father.

"Can we?" said Rollo; "I don't see how."

"We cannot see the air, it is true; but then we can see its effects, and so we can experiment upon it."

"Well, at any rate," said Rollo, "we can't build a dam, and make it spout through a hole, like water."

"No," said his father, "not exactly. In your dam, for instance, when it was full, you had water on one side of the board, and no water on the other; and then, by opening a hole in the board, the water spouted through; but we cannot very well get air on one side of a partition, and no air on the other; if we could, it would spout through very much as the water did."

"Why can't we do that, sir?" said Rollo.

"Because," replied his father, "we are all surrounded and enveloped with air. It spreads in every direction all around us, and rises many miles above us. Whereas, in respect to water, you had one little stream before you, which you could manage just as you pleased. If you were down at the bottom of the sea, then the water would be all around you and above you; and there, even if you could live there, you could not have a dam."

"No, sir," said Rollo, "the water would be everywhere."

"Yes," replied his father, "and the air is everywhere. If, however, we could get it away from any place, as, for instance, from this room, then bore a hole through the wall, the weight of the air outside would crowd a portion of it through the hole, exactly as the weight of the water above the board in your dam crowded a part through the hole in the board."

"I wish we could try it," said Rollo.

"We *can* try it, in substance," said his father, "in this room; or—no, the china closet will be better."

There was a china closet, which had two doors in it. One door opened into the parlor, where Rollo and his father were sitting. The other door opened into the back part of the entry. Rollo's father explained how he was going to perform the experiment, thus:—

"If we could, by any means, get all the air out of the closet for a moment, then the pressure of the air outside would force a jet of it in through the key-holes of the doors, and the crevices."

"And how can we get the air out?" said Rollo.

"We can't," said his father, "get it all out; but we can get a part of it out by shutting the door quick. The door will carry with it a part of the air that was in the closet, and then the outside air will be spouted in, through the key-hole of the other door. Only we can't see it, as we can the water."

"No," said Rollo; "but I can put my hand there, and feel it."

"A better way," said his father, "would be to hold a lamp opposite to the key-hole, and see if it blows the flame."

Rollo tried the experiment, in the way his father had described. He went into the closet with the lamp. He held the lamp opposite to the key-hole, and pretty near to it, and then he asked Nathan to shut the other door suddenly. Nathan, who was standing all ready by the other door, which was about half open, put his two hands against it, and pushed it to, with all his strength, producing a great concussion.

"O Nathan," said his father, "you need not be quite so violent as that."

"It succeeded, father, it succeeded," said Rollo.

"I'm glad it succeeded," said his father; "but Nathan need not have shut the door with so much force."

"I wanted to drive out all the air," said Nathan.

"I'll show you how to do it," said his father.

Rollo's father accordingly arose, and came to the closet door. He opened the door wide, and then explained to the boys, that the beginning of the movement of the door, when it was wide open, did not drive out any air.

"For," said he, "there is so large a space between the edge of the door and the wall, that the air that is put in motion by the movement of the door, can pass directly round the edge, back into the closet again. It is only when the door is almost shut, when the edge of it comes close to the casing all around, that the movement of the door drives the air out."

Then he took hold of the latch of the door, and put it almost to, very gently. He turned the latch so as to prevent its snapping against the catch, and then pushed it suddenly into its place three or four times, opening the door only a very little way every time.

"Now," said he, "hold the lamp at the key-hole, and watch the flame, while I shut the door two or three times in this way."

Rollo did so, Nathan standing all the time by his side. They observed that the flame of the lamp was driven into the room every time the door was shut; proving that, every time a little of the air was driven out by the door, a little puff rushed in at the key-hole.

"Let us stop up the key-hole," said Rollo, "and then it can't get in."

"Yes," said his father, "there are a great many little crevices all around the closet, where the air can come in."

"Couldn't we stop those up too?" said Rollo.

"No," said his father, "not so as to make the closet air-tight. For, if the crevices could all be stopped exactly, the air would come in through the very wood itself."

"How?" said Rollo.

"Why, there are little pores in wood, that is, little channels that the sap flowed in when the wood was growing, and the air can pass through these."

Here Rollo's father observed that Rollo was looking very intently at the table; and he asked him what he was doing: he said he was trying to find some of the pores.

"You can't see them there," said his father. "St. Domingo mahogany is a very hard and close-grained kind of wood. If it was summer, and you could dig down and get a small piece of the root of the great elm-tree in the yard, you could see the pores and channels there."

After some more conversation on this subject, Rollo asked his father if he could not think of some other experiments for them to try. His father said that he did not just then think of any experiment, but that, if Rollo and Nathan would come and sit down by the fire, he would give them some information on the subject. Rollo's mother said that she should like to hear too. They accordingly waited until she was ready, and then, when all were seated, Mr. Holiday began thus:—

"Air is in many respects much like water."

"Yes," interrupted Rollo, "just like water, only thinner, because, you see——"

"You must not interrupt me," said his father, "unless to ask some question, which is necessary to understand what I say. It is entirely irregular for a pupil, instead of listening to his teacher, to interrupt, in order to tell something that he knows himself."

Rollo's father smiled, as he said this, but Rollo looked rather ashamed. Then his father proceeded: —

"There is one very remarkable difference between them. Water is not compressible by force; but air is."

"What is the meaning of *compressible*?" said Nathan.

"Compressible things," said his father, "are those that can be compressed, that is, pressed together, so as to take up less room than they did before. Sponge is compressible. A pillow is compressible. But iron is not compressible, and water is not compressible."

"I should think it was," said Nathan; "it is very soft."

"It is very *yielding*," replied his father, "when you press it, but it is not pressed into any smaller space. It only moves away. If you have a tumbler half full of water, and press a ball down into it, you could not crowd the water into any smaller space than it occupied at first; but, as fast as the ball went down, the water would come up around the sides of the ball."

"But suppose," said Rollo, "that the ball was just big enough to fit the tumbler all around; then the water could not come up."

"And then," said his father, "you could not crowd the ball down."

"Could not a *very* strong man?" said Nathan.

"No," replied his father, "the water cannot be sensibly compressed. But now, if the tumbler contained only air, and if a ball were to be put in at the top, just large enough to fit the tumbler exactly, and if a strong man were to crowd it down with all his strength, he would, perhaps, compress the air into half the space which it occupied before."

"Perhaps the tumbler would break," said Nathan.

"Yes," replied his father, "and the tumbler will answer only for a supposition; but for a real experiment it would be best to have a cylinder of iron."

"What is a cylinder?" said Nathan.

"An iron vessel, shaped like a tumbler, only as large at the bottom as it is at the top, would be a cylinder. Now, if there was a cylinder of iron, with the inside turned perfectly true, and a brass piston fitted to it——"

"What is a piston?" said Nathan.

"A piston," said his father, "is a sort of stopper, exactly fitted to the inside of a cylinder, so as to slide up and down. It is made to fit perfectly, and then it is oiled, so as to go up and down without much friction, that is, hard rubbing. There is a sort of stem coming up from the middle of the piston, called the piston rod, which is to draw up the piston, and to press it down by.

"Now," continued his father, "if a strong man had a cylinder like this, with a piston fitted to it, and a strong handle across the top of the piston rod, perhaps he might press the air into one half the space which it occupied before. That is, if the cylinder was full of air when he put the piston in, perhaps he could get the piston down half way to the bottom. Then the air would be twice as *dense* as it was before; that is, there would be twice as much of it in the same space as there was before. It would be twice as compact and heavy. This is called *condensing* air. The philosophers have ingenious instruments for condensing air.

"If, however, a man condenses air in this way, by crowding down a piston, he does not begin the condensation when the piston begins to descend. The air is condensed a great deal before he begins. All the air around us is condensed."

"How comes it condensed?" said Rollo.

"Why, you recollect that, when you bored a hole through the board in the bottom of your dam, the water spouted out."

"No, father," said Rollo, "we pulled the plug out; Jonas bored the hole."

"Well," said his father, "the water spouted out."

"Yes," said Rollo.

"What made it?" said his father.

"Why, the water above it was heavy, and pressed down upon it, and crowded it out through the hole."

"Yes," said his father, "and the deeper the water, the more heavily it was pressed."

"Yes, sir," said Rollo, "and the farther it spouted."

"Because it was pressed down by the load of such a high column of water."

"Yes, sir," said Rollo.

"Well," replied his father, "it is just so with the air. The air all around us is pressed down by the load of all that is above us. We are, in fact, down at the bottom of a great ocean of air, and the air here is loaded down very heavy."

"How heavy?" said Rollo.

"O, very heavy indeed," said his father.

"Why, air is pretty light," said Rollo.

"Yes," replied his father, "but then the column of it is very high."

"How high?" said Rollo.

"Why, between thirty and forty miles. But it grows thinner and thinner towards the top; so it is not as heavy, by any means, as a column of air would be, thirty miles high, and as dense all the way up as it is here."

"What makes it grow thinner and thinner towards the top?" said Rollo.

"Because," said his father, "that which is near the top, has not as much load of air above it, to press it down."

"And that which is *at* the top," said Rollo, "has none above it, to press it down."

"No," replied his father.

"And how thin is it there?"

"Nobody knows," said his father.

"What, nobody at all?" said Nathan.

"No, I believe not; at least I do not; and I don't know that any body does."

"How do they know, then, how high it is?" said Rollo.

"The philosophers have calculated in some way or other, though I don't exactly know how. I believe they have ascertained how great the pressure of the air is here at the surface of the earth, and have calculated in some way, from that, how high the air must be to produce such a pressure."

"And how high must it be?" said Nathan.

"Why, between thirty and forty miles," said Rollo; "father told us once."

"And yet," continued his father, "water, thirty or forty feet deep, would produce as great a pressure as a column of air of thirty or forty miles. That is, the air around presses about as heavily, and would force a jet of air through a hole with about as much force, as water would, coming out at the bottom of a dam, as high as a common three-story house."

These explanations were all very interesting to Rollo and to his mother; but Nathan found it rather hard to understand them all, and he began to be somewhat restless and uneasy. At length he said,—

"And now, father, haven't you almost done telling about the air?"

"Why, yes," said his father; "I have told you enough for this time; only you must remember it all."

"I don't think I can remember it quite all," said Nathan.

"Well, then, remember the general principle, at any rate," said his father, "which is this—that we live at the bottom of a vast ocean of air, and that the lower portions of this air are pressed down by the load of all the air above; that, being so pressed, the lower air is condensed,—so that we live in the midst of air that is pressed down, and condensed, by the load of all that is above it; and that, consequently, whenever the air is taken away, even in part, from any place, as you removed some of it from the china closet, the pressure upon the air outside forces the air in through every opening it can find."

"I think that is a little too much for me to remember," said Nathan.

Nathan's father and mother laughed on hearing this, though Nathan did not know what they were laughing at. His father told him that he could not expect him to remember all; and that, to pay him for his particular attention, he would tell him a story.

So he took Nathan up in his lap, and told him a very curious story of a boy, who went about the yard with a little dog upon one of his shoulders, a cat upon the other, and a squirrel on his head. The squirrel was tame.

QUESTIONS

Why cannot experiments be performed upon the pressure of air, as conveniently as upon the pressure of water? How did Rollo's father contrive to remove a part of the air from the china closet? Where did they expect that the air would be forced into the closet? How were they to make this effect visible? Did the experiment succeed? Suppose the key-hole had been stopped up; where would the air have been forced in? Suppose all the crevices had been closed. Is water compressible? Is air compressible? What is the shape of a cylinder? What is a piston? How might air be compressed by means of a cylinder and piston? What was the general principle which Rollo's father stated, in conclusion?

CHAPTER VI
BALLOONING

THE next evening, Rollo and Nathan had another conversation with their father, respecting air. When they were all seated, he commenced as follows:—

"I told you yesterday, that air may be compressed by force, while water cannot be. It has another property, which is in some respects the reverse of this. It springs back into its original bulk, when the pressure is removed."

"How?" said Nathan; "I don't exactly understand you."

"Why, you remember what I said about the experiment with the iron cylinder and a piston to fit it."

"Yes, sir," said Rollo.

"What was the experiment?" said his father.

"Why, if a man were to press the piston down hard, he could crowd the air all into the lower half of the cylinder."

"Yes," replied his father. "Now, the property I am going to tell you about this evening is this—that, if the man lets go of the piston rod, the air that is condensed into the bottom of the cylinder, will spring up, and force the piston up again. This property is called *elasticity*. It is sometimes called the *expansive force* of the air. For it is a force tending to expand the air, that is, to swell it out into its original dimensions. This is another great difference between air and water.

"Now, as all the air around us," continued Rollo's father, "is pressed down very heavily, and is condensed a great deal, it is all the time endeavoring to expand; and it would expand, were it not that the great burden of the air above it keeps it condensed. But water is not compressed, and has no tendency to expand. The water of Rollo's dam, for instance, had all the weight of the atmosphere resting upon it, but it did not compress it at all, and so it did not tend to expand.

"And now," said his father, "I cannot perform any experiment, to show you that air tends strongly to expand or swell out into a great space, while water does not; but I can make a supposition, which will illustrate it. Suppose we had a large, but very thin, glass bottle, filled with water, and put down upon the floor in the middle of this room. Suppose, also, that we had another bottle, of the same size and shape, filled with air, and we put that down upon the floor by the side of the other; both bottles being stopped very tight. Now, if we could by any means suddenly take away all the air from the room, so that there should be nothing around the bottles, then the bottle of water would remain just as it is, for the glass would have nothing to support but the weight of the water, and it would be strong enough for that. But the bottle of air would fly all to pieces; for that would not rest quietly, like the water, satisfied with the space which it already has, and only pressing with its own weight upon the sides of the glass; but it would immediately expand with so much force as to break the thin glass all to pieces."

"Would it!" exclaimed Rollo and Nathan together. "And would it make a loud noise?"

"Yes," replied their father, "I presume it would make a loud explosion; that is, if the air in the room around it could by any means be all at once and suddenly removed.

"And so you must remember," he continued, "that there are two very remarkable differences between air and water. Air may be condensed by pressure, and, as it exists all around us, is greatly condensed by the pressure of the air above, and it may be compressed more. And air is expansive, while water is not. Whenever the pressure upon it is removed, it suddenly expands, or spreads out in all directions."

"O dear me!" said Nathan, with a sigh.

"What is the matter?" said his father.

"Why, I can't understand it very well."

"Can't you?" said his father. "Well, I must admit that you are rather too young to study pneumatics."

"Pneumatics?" repeated Rollo.

"Yes," said his father; "that is the name of this science."

—**"Then it sailed slowly away"**—

"What, the science of air?" said Rollo.

"Yes," said his father, "the science which treats of air, and of all other compressible and expansive fluids. But let me think. I must try to tell you something which Nathan can understand and be interested in. If I had a very light feather, I could let him perform an experiment."

"Would a little down do?" said Rollo's mother.

"Yes," replied his father, "that would be better than a feather."

Mrs. Holiday then went and brought a little down, and handed it to Rollo's father. Now, there was a lamp upon the table, of a peculiar kind, called a study lamp. It had a glass tube, called a chimney, around the wick, and consequently around the flame itself, being round, like a ring.

Rollo's father told Nathan to hold the down over the top of this glass chimney, and then to let it go.

Nathan did so. The little tuft of down was wafted up into the air, quite high above the lamp, and then it sailed slowly away, and fell down upon the table.

"I know what makes it rise," said Rollo. "It is the heat. The heat makes it rise."

"Do you think so?" said his father. "Then take the down, and lay it gently upon the hearth, before the fire, as near as you can."

Rollo did so. He had to take his hand away very quick, for it was quite hot there. The little tuft remained quietly upon the hearth where he placed it.

"There," said his father, "is not that a hotter place than it was over the lamp?"

"Yes, sir," said Rollo.

"Then, if it was heat that made it rise, why does not it rise now?"

Rollo could not tell.

"I will tell you how it was," said his father. "Heat makes air more expansive. When air is heated, it swells; when it is cool, it shrinks again. Now, if it swells, it becomes lighter, and so it is buoyed up by the heavier air around it; just as wood at the bottom of the sea would be buoyed up, and would rise to the surface of the water. Now, the heat of the lamp heats the air that is in the glass chimney, and swells it. This makes it lighter; and so the air around it, which is heavier, buoys it up, and it carries up the feather with it."

"No, the down, father," said Nathan.

"Yes, the down," said his father.

"Then it seems to me, after all," said Rollo, "that it is the heat which makes it rise."

"Yes," said his father, "it does, indirectly. It expands the air; that makes it lighter; then the heavy air around it buoys it up, and, when it goes up, it carries up the down. So that it is not strictly correct to say, that the heat carries it up. The heat sets in operation a train of causes and effects, the last of which results in carrying up the feather.

"Now," continued his father, "there is always a stream of air going up, wherever there is a lamp, or a fire, or heat, which heats the air in any way. The expanded air from a fire goes up chimney. The cool and heavy air in the room and out of doors crowds it up."

"The air out of doors?" said Rollo. "How can that crowd it up?"

"Why, it presses in through all the crevices and openings all around the room, and crowds the light air up the chimney. All the smoke is carried up too with it, and it comes pouring out at the top of the chimney all the time."

"You can see that the air presses in at all these crevices," continued Rollo's father, "by experiment."

"What experiment is it?" said Rollo; "let us try it."

"I will let Nathan try it," said his father, "and you may go with him and see the effect. First," he continued, "you see by the smoke, that the air really goes up the chimney; and I will show you that other air really crowds into the space, from other parts of the room."

So he took a lamp from the table,—not the study lamp; it was a common lamp,—and held it at various places in the opening of the fireplace, by the jambs and near the upper part; and Rollo and Nathan saw that the flame, in all cases, was turned in towards the chimney.

"Yes," said Rollo, "I see it is drawn in."

"No," said his father; "strictly speaking, it is not *drawn* in; it is pressed in, by the cool and heavy air of the room."

"I thought," said Rollo's mother, "that the chimney *drew* the air from the room into it."

"That is what is generally said," replied Mr. Holiday, "but it is not strictly true. The common idea is, that the hot air rises in the chimney, and so draws the air from the room to supply its place; but this is not so. In the first place, nothing can rise unless it is forced up. The lightest things have some weight, and would, if left to themselves, fall. The hottest and lightest air in a chimney would fall to the earth, if there was no cooler and heavier air around it, to force it to rise;—just as the lightest cork, which would rise very quick from the bottom of the sea, would fall back again very quick, if the water was not there.

"Remember, then, Nathan and Rollo, that, when a fire is built in a fireplace, so as to warm the air in the chimney, it makes this air not so heavy; and then the cool air all around it in the room and out of doors, presses in, and crowds under the light air, and makes it ascend."

"But, father," said Nathan, "you said I might perform an experiment."

"Very well, I am ready now. Take the lamp, and carry it around the room, and hold it opposite any little opening you can find."

"I can't find any little openings," said Nathan.

"O yes," said his father; "the key-hole of the door is a little opening, and there is a narrow crevice all around the door; and you will find little crevices around the windows. Now, hold the lamp opposite any of these, and you will see that the air presses in."

So Nathan went with the lamp, Rollo following him, and held the lamp opposite to the key-hole, and the crevices around the door and windows; only, when he came to the window, his father told him to be very careful not to set the curtain on fire.

Rollo wanted Nathan to let him try it once; and so Nathan gave him the lamp. He said he meant to make a crevice; and so he pushed up the window a very little way, and held the lamp opposite to the opening. The air pressed the flame in towards the room, in all cases.

"People commonly say, that it is *drawn* in," said his father, "but that is not strictly correct; it is really *pressed* in. There is no power of attraction, in the air that is in the room, to draw in the air that is out of doors, through the crevices; but the air that is out of doors, is so heavy, that it presses in, and crowds the warm and light air up the chimney.

"And now," said his father, "I cannot tell you anything more this evening; but, if you remember this, I will give you some further instruction another time."

"Well, sir," said Nathan, "only I wish you would tell me a little story, as you did last evening. Have not I been still?"

His father had noticed, that he had been very still and attentive, but did not think before, that it was in expectation of being rewarded with a story.

"Well," said his father, "I will tell you a story, or give you a little advice. How should you like a little advice?"

"Well, father, a little advice; just which you please."

"I advise you, then,—let me see,—what shall I advise you?—No, on the whole, I will tell you a story. Once there was a man, and he was a philosopher. He understood all that I have been explaining to you about the air being light when it was hot. So he got some very thin paper, and made a large paper bag. He cut the paper very curiously, and pasted it together at the edges in such a way, that the bag, when it was done, was round, like a ball; and it had a round opening at the bottom of it. In fact, it was a large paper ball."

"How large was it?" said Nathan.

"It was so large, that, when it was swelled out full, it would have been higher than your head."

"O, what a large ball!" said Nathan. "But what was it for?"

"Why, the man thought, as hot air is lighter than cool air, and floats up, that perhaps, if he could fill his paper ball with hot air, it would go up too."

"And did it?" said Nathan.

"Yes," said his father. "He filled it with hot air; and the hot air was so light, that it rose up and carried the paper ball with it."

"How did he get the hot air into it?" said Rollo.

"Why, he held it over a little fire, with the mouth down. Then the hot air from the fire went up into the ball, and swelled it out full."

"How high did it go?" said Nathan.

"O, it soared away," said his father, "away up into the air, very high; until at length it got cool, and then it came down."

"I should like to see such a ball as that," said Nathan.

"Such a ball as that is called a *balloon*," said his father.

"I wish I could see a balloon," said Nathan.

QUESTIONS

What is the important difference between air and water, which was explained in the last chapter, and mentioned in this? Does the air tend to expand again after it is compressed? What is this property of the air called? Is the air around us already condensed, or is it in its natural state? What causes it to be condensed? Suppose a thin glass vessel were to be filled with air, and another with water, and the air suddenly removed from the room around them; what would be the effects? What effect does heat have upon the expansibility of air? How may this effect be made to appear over a lamp? In a chimney? What was the story which Rollo's father told Nathan?

CHAPTER VII
PHILOSOPHICAL DISCUSSION

SOME time after this, Rollo, and Nathan, and James, were playing in the shed, one pleasant morning in the spring. They were building with sticks of wood, which they piled in various ways, so as to make houses. They took care not to pile the wood, in any case, higher than their shoulders, for Jonas had told them that, if they piled the wood higher than that, there would be danger of its falling down upon them.

After some time, Rollo went into the house a few minutes, and James and Nathan went to the open part of the shed, and began to look out of doors. The sun was shining pleasantly, but the ground was wet, being covered with streams and pools of water, and melting snow-banks.

"What a pleasant day!" said James. "I wish it was dry, so that we could go out better."

"I wish we could fly," said Nathan, "for it is very pleasant up in the air."

"I wish we had a balloon," said James. "If we had a balloon, we could go up in the air, easier than to fly."

"O James," said Nathan, "you could not get into a balloon if you had one."

"Why not?" said James.

"Because," said Nathan, "it would not be big enough."

"Why, Nathan," said James, "a balloon is bigger than this house."

"O James, it is not higher than my head."

"It is," said James, "I know it is. I have read about balloons bigger than a house."

"And my father," said Nathan, putting down his foot in a very positive air, "my father told me himself, that a balloon was about as high as my head."

The boys disputed some time longer upon the subject, and finally, when Rollo came out of the house, they both appealed to him very eagerly to settle the dispute.

"Isn't a balloon higher than Nathan's head?" said James.

"Is it as high as a house?" said Nathan.

"Why, I know," said Rollo, "that a man made a balloon once about as high as Nathan's head, because my father said so; but perhaps it was a little one."

"Yes," said James, "I know it must be a little one; for balloons are big enough for men to go up in them."

"O James," said Nathan, "I don't believe it. Besides, the fire would burn 'em."

"What fire?" said James.

"The fire they burn under the balloons, to make the air hot," said Nathan.

"I don't believe they have any fire," said James.

Just then Nathan, happening to look around, saw Jonas standing behind them; he had just come out of the house, and was going out to his work. Hearing the boys engaged in this dispute, he stopped to listen. The boys both appealed to Jonas.

Jonas heard all that they had to say, and then replied,—

"I cannot tell you much about going up in a balloon, but I can tell you something about getting along pleasantly down here upon the earth, which I think may be of service to you."

"What is it?" said James.

"Why, that you will neither of you get along very pleasantly until you can bear to have any body else mistaken, without contradicting them. James, you think Nathan is mistaken about the size of a balloon, do you?"

"Yes, I know he is," said James.

"Well," said Jonas, "now why not let him remain mistaken?"

"Why,—I don't know," said James.

"He isn't willing to be convinced, is he, that a balloon is as big as a house?"

"No," said James, "he is not."

"Then why don't you let him remain unconvinced? Why should you insist on setting him right, when he don't want to be set right?"

"And you, Nathan, suppose that James is mistaken, in thinking that the balloon is so big."

"Yes," said Nathan, "and that men can get into it, and go up in the air."

"Well, now, if he wants to believe that balloons are so big, why are you not willing that he should? Why should you insist upon it that he should know that they are smaller?"

"Because I *know*," said Nathan, very positively, "that they are small; and, besides, the paper would not be strong enough to bear a man."

"I did not ask you," said Jonas, "why *you* believed that men could not go up in balloons, but why you were so anxious to make James believe so. Why not let him be mistaken?"

"Why—because," said Nathan.

"You see, Nathan," continued Jonas, "the world is full of people that are continually mistaken; and if you go about trying to set them all right by disputing them, you'll have a hard row to hoe."

"A hard what?" said Nathan.

"A hard row to hoe," repeated Jonas. "It's never of any service to attempt to convince people that don't want to be convinced; especially if they are wrong."

"Especially if they are wrong!" repeated Rollo, in astonishment.

"Yes," replied Jonas. "The very worst time to argue with a boy, is when he is wrong, and does not want to be set right. It is a great deal harder to get along in argument with one who is right, than with one who is wrong; for the one who is wrong, disputes; the one who is right, reasons."

"Well, Jonas," said James, "which of us was disputing?"

"Both of you," said Jonas.

"Both of us," said James; "but you said that only the one who was wrong, disputed."

"Well," replied Jonas, "you were both wrong."

"Both wrong! O Jonas!" said James.

"Yes, both wrong," replied Jonas; and so saying, he was going away to his work.

"But stop a minute longer," said James, "and tell us how it is about the balloon; we want to know."

"O no," said Jonas, "you don't want to know; you want to *conquer.*"

"What do you mean by that?" said Nathan.

"Why, you don't really wish to learn any thing; but you want to have me decide the case, because each of you hopes that I shall decide in his favor. You want the pleasure of a victory, not the pleasure of acquiring knowledge."

"No, Jonas," said Nathan, "we do really want to know."

"Well," said Jonas, "I can't stop now to tell you; perhaps I will this evening; but I advise you always, after this, not to contradict people, and dispute with them, when they say things that you don't believe. Do as the gentleman did, when the man said the moon was a fire."

"What did he do?" said Rollo.

"Why, he let him say it as much as he wanted to."

"Tell us all about it," said James.

"Well, then," said Jonas, "once there was a man, and he saw the moon coming up behind the trees, and thought it was a large house burning up. He went along a little way, and he met a vulgar fellow, riding in a carriage."

"Riding in a carriage!" repeated Rollo, astonished.

"Yes," said Jonas, "handsomely dressed. 'Sir,' said the man, 'see that great fire!'

"'It isn't a fire, you fool,' said the vulgar fellow; 'it's nothing but the moon.'

"'The moon!no it isn't,' said the man; 'it is a monstrous great fire. Don't you see how it blazes up?'

"'It is not a fire, I tell you,' said the vulgar fellow.

"'I tell you 'tis,' said the man.

"'You don't know any thing about it,' said the vulgar fellow.

"'And you don't know the moon from a house on fire,' said the man, and so passed on.

"A minute or two after this, he met a gentleman driving a team."

"A gentleman driving a team!" said James.

"Yes," said Jonas, "with a frock on. He was tired and weary, having driven all day. The man asked him if he did not see that house on fire.

"'Ah,' said the gentleman, 'I thought it was the moon.'

"'No,' said the man, 'it is a house on fire.'

"'Well,' said the gentleman, 'I am very sorry if it is. I hope they'll be able to put it out!'

"So saying, he started his team along, and bade the man good evening."

Jonas then, having finished his story, stepped out of the shed, and went along towards the barn; Nathan called out after him to say,—

"Well, Jonas, I don't understand how the gentleman came to be driving a team all day."

Jonas did not reply to this, but only began to laugh heartily, and to walk on. Nathan turned back into the shed, saying, he did not see what Jonas was laughing at.

The boys wanted very much to have the question about the balloon settled; and, after some further conversation on the subject, they concluded to go in and ask their mother. So they all three went in. Rollo proposed this plan, and he led the way into the house. He found his mother sitting in the parlor at her work.

"Well, boys," said she, "have you got tired of your play?"

"No, mother," said Rollo, "but we want to know about balloons: how big are they?"

"O, some of them," said she, "are very large."

"Ain't they as big as this house?" said James.

"Yes, I believe they have been made as big," said she.

"But, mother," said Nathan, "father told me, his very self, that they were no higher than my head."

"O no," said his mother; "he said that a man made *one* which was about as high as your head; but that was only a little one, for experiment. When they make large ones, for use, they are as high as this house."

"For use, mother?—what use?" said Nathan.

"Men go up in them, don't they, aunt?" said James.

"Not *in* them, exactly," said his aunt. "They could not live in them, but they go up *by means* of them."

"How?" said Nathan.

"Why, they have a kind of basket, which hangs down below the balloon, and they get into that."

"I knew they could not get into the balloon," said Nathan.

"Then you have had a dispute about it," said his mother.

"Why,—yes," said Nathan, with hesitation, "we disputed a little."

"I am sorry to hear that," said his mother, "for disputing seldom does any good. The fact is, however, that men have often been carried up by balloons, but they never get into them. They could not live in them. They could not breathe the kind of air which balloons are filled with."

"It is hot air," said Nathan.

"No," said his mother, "the kind of balloon which your father told you of was filled with hot air; but the balloons which people generally use to go up with, are filled with another kind of air, which is very light when it is cool. They make an enormous bag of silk, and fill it with this light air, which they make in barrels; and then, when the bag is filled, it floats away above their heads, and pulls hard upon the fastening. There is a net all over it, and the ends of the net are drawn together below, and are fastened to the basket, or car, where the man is to sit. When it is all ready, the man gets into the car, and then they let go the fastenings, and away the great bag goes, and carries the man with it, away up into the air."

"And then how does he get down?" said Nathan.

"Why, he can open a hole in the bag, and let some of the light air out; and then he begins to come down slowly. If he comes down too fast, or if he finds that he is coming into the water, or down upon any dangerous place, there is a way by which he can make his balloon go up again."

"What way is it, aunt?" said James.

"Why, he has some bags of sand in his balloon," said his aunt; "and the balloon is made large enough to carry him and his sand-bags too. Then, if he finds that he is coming down too fast, he just pours out some of his sand, and that makes his car lighter; and so the balloon will carry him up again."

"That's a good plan," said Rollo.

"Yes," said his mother; "the reason why he takes sand is, because that will not hurt any body by falling upon them. If he should take stones, or any other heavy, solid things, and should drop them out of his car, they might possibly fall upon some body, and hurt them. So he takes sand in bags, and, when he wants to lighten his balloon, he just pours the sand out."

Rollo's mother then told the boys that there was a large book, which had several stories in it of men's going up in balloons, and that she would get it for them. So she left her work, and went out of the room; but in a few minutes she returned, bringing with her two very large, square books, with blue covers. One of them had pictures in it, and among the rest there were pictures of balloons. She opened the other book, and found the place where there was an account of balloons, and she showed the place to Rollo.

She told the boys that they had better go out in the kitchen, or into the shed, if it was warm enough, and read the account.

"You and James, Rollo," said she, "can read by turns, and let Nathan hear. Then, when the plates are referred to, you must look into the other book and find them."

"Yes," said Rollo, "we will; only, mother, if you would let us sit down here and read it—and then, if there is any thing which we cannot understand, you can tell us what it means."

"Very well," replied his mother, "you may sit down here upon the sofa."

So the boys sat down upon the sofa. They put Nathan between them, so that he might look over. Rollo and James took turns to read, and they continued reading about balloons for more than an hour. There was one story of a sheep, which a man carried up in his car, under a balloon, and then let him drop, from a great height, with a parachute over his head, to make him fall gently. And he did fall gently. He came down to the ground without being hurt at all.

QUESTIONS

How was the subject of balloons introduced into the conversation? What was Nathan's opinion about the possibility of being carried up by a balloon? What was the dispute about the size of balloons? What was Nathan's evidence? What was James's evidence? What did Jonas say when they appealed to him? What was the story that he related? Which of the boys did he finally say was wrong? Whom did the boys appeal to afterwards? What did Rollo's mother say about the size of balloons? How did she say that large balloons were filled? How can they make the balloon come down? How can they make it go up again, if they wish to do so?

CHAPTER VIII
TASKS

A FEW days after this, there commenced a long storm of rain. Rollo and Nathan were glad to see it on one account, for their mother told them it would melt away the snow, and bring on the spring. The first day, they amused themselves pretty well, during their play hours, in the shed and in the garret; but on the second day, they began to be tired. Nathan came two or three times to his mother, to ask her what he should do; and Rollo himself, though, being older, his resources might naturally be expected to be greater, seemed to be out of employment.

At last, their mother proposed that they should come and sit down by her, and she would tell them something more about the air. "How should you like that, Rollo?" said she.

"Why, pretty well," said Rollo; but he spoke in an indifferent and hesitating manner, which showed that he did not feel much interest in his mother's proposal.

"*I* can't understand very well about the air," said Nathan.

Their mother, finding that the boys did not wish much to hear any conversation about the air, said nothing more about it just then, and Rollo and Nathan got some books, and began to read; but somehow or other, they did not find the books very interesting, and Rollo, after reading a little while, put down his book, and went to the window, saying that he wished it would stop raining. Nathan followed him, and they both looked out of the window with a weary and disconsolate air.

Their mother looked at them, and then said to herself, "They have not energy and decision enough to set themselves about something useful, and in fact I ought not to expect that they should have. I must supply the want, by my energy and decision."

Then she said aloud to Rollo and Nathan,—

"I want you, boys, to go up into the garret, and under the sky-light you will see a large box. Open this box, and you will find it filled with feathers. Select from these feathers three or four which are the most downy and soft about the stem, and bring them down to me."

"What are they for?" said Rollo.

"I will tell you," replied his mother, "when you have brought them to me."

So Rollo and Nathan went up into the garret, and brought the feathers. They carried them to their mother. She said that they would answer very well, and she laid them gently down upon the table.

Then she took up her scissors, and began to cut off some of the lightest down, saying, at the same time,—

"Now, children, I am going to give you some writing to do, about *the air.*"

"Writing?" said Rollo.

"Yes," said his mother. "I am going to explain to you something about the air, and then you must write down what I tell you."

"But I can't write," said Nathan.

"No," said his mother, "but you can tell Rollo what you would wish to say, and he will write it for you."

"Why, mother," said Rollo, "I don't think that that will be very good play."

"No," replied his mother, "I don't give it to you for play. It will be quite hard work. I hope you will take hold of it energetically, and do it well.

"First," said she, "I am going to perform some experiments for you, before I tell you what I want you to write."

By this time, she had cut off the downy part of several feathers, and had laid them together in a little heap. Then she took a fine thread, and tied this little tuft of down to the end of it. Then she took up the thread by the other end, and handed it to Rollo.

"There, Rollo," said she. "Now, do you remember what your father told you, the other day, about the effect of heat upon air?"

"It makes it light," said Rollo.

"And why does it make it light?" asked his mother.

"Why, I don't exactly recollect," said Rollo.

"Because it swells it; it makes it expand; so that the same quantity of air spreads over a greater space; and this makes it lighter, But cool or cold air is heavier, because it is more condensed.

"Now, wherever there is heat," continued his mother, "the air is made lighter, and the cool and heavy air around presses in under it, and buoys it up. This produces currents of air. You recollect, don't you, that your father explained all this to you the other day?"

"Yes," said Rollo, "I remember it."

"Well," said his mother, "now you and Nathan may take this little tuft, and carry it about to various places, and hold it up by its thread, and it will show you the way the air is moving; and then you may come to me, and I will explain to you why it moves that way."

"Well," said Rollo, "come, Nathan, let us go. First we will hold it at the key-hole of the door."

Rollo held the end of the thread up opposite to the door, in such a way, that the tuft was exactly before the key-hole. The tuft was at once blown out into the room.

"O, see, Nathan, how it blows out. The air is coming in through the key-hole."

"Yes," said his mother; "when there is a fire in the room, and none in the entry, then the cold air in the entry runs down through the key-hole into the room."

"It don't run down, mother," said Rollo; "it blows right in straight."

"Perhaps I ought to have said it spouts in," said his mother, "just as the water did from the hole in your dam. And, now," she continued, "come and hold the tuft near the chimney."

Rollo did so; and he found that it was carried in, proving, as their father had showed them before, that the heavy, cold air, pressing into the room, crowded the warm, light air up the chimney.

"Now, should you think," said their mother, "that the cold air could come in through the key-hole, as fast as it goes up the chimney?"

Both Rollo and Nathan thought that it could not.

"Then go all around the room," said she, "and see if you can find any other place, where it comes in. For it is plain, you see, that the light air cannot be driven up chimney any faster than cold and heavy air comes in to drive it up and take its place."

So Rollo and Nathan went around the room, holding their tuft at all the places they could find, where they supposed there could be openings for the cold air to press in. They found currents coming in around the windows, and by the hinges of the doors; and at length Rollo said, he meant to open the window a little way, and see if the cold air from out of doors would not press in there too. He did so, and the tuft was blown in very far, showing that the cold air from out of doors pressed in very strongly.

"Now, if all these openings were to be stopped," said their mother, "then no cold air could crowd into the room; and of course the hot air could not be buoyed up into the chimney, and a great deal of the hot air and smoke would come into the room. This very often happens when houses are first built, and the rooms are very tight.

"But now, Rollo," she continued, "suppose that the door was opened wide; then should not you think that *more* cold and heavy air would press in, than could go up the chimney?"

"Yes, mother, a great deal more," said Rollo.

"Try it," said his mother.

So Rollo opened the door, and held his tuft in the passage-way; and he found that the air was pressing in very strongly through the open space. Wherever he held it, it was blown into the room a great deal, showing that the heavy air pressed in, in a torrent.

"Now, as much warm air must go out," said she, "as there is cold air coming in; but I don't believe that you and Rollo can find out where it goes out."

Rollo looked all around the room, but he could not see any opening, except the chimney and the door, and the little crevices, which he had observed about the finishing of the room. He said he could not find any place.

His mother then told him to hold his tuft down near the bottom of the doorway. He did so, and found that the current of air was there very strong. The tuft swung into the room very far.

"Now hold it up a little higher," said his mother.

Rollo obeyed, and he found that it was still pressed in, but not so hard.

"Higher," said his mother.

Rollo raised it as high as he could reach. The thread was of such a length, that the tuft hung about opposite to his shoulder. The tuft was still pressed in, but not nearly as far as before.

"So you see," said his mother, "that the air pours in the fastest at the lowest point, where the weight and pressure of the air above it are the greatest; just as, in your dam, the water from the lowest holes spouted out the farthest."

"Yes," said Rollo, "it is very much like that."

"Now," continued his mother, "you see that a great deal of air comes in, and if you look up chimney, you will see that there is scarcely room for so much to go up there;—and yet just as much must go out as comes in.

"Get the step-ladder," said his mother, "and stand up upon it, and so hold your tuft in the upper part of the door-way."

There was in the china closet a small piece of furniture, very convenient about a house, called a step-ladder. It consisted of two wooden steps, and was made and kept there to stand upon, in order to reach the high shelves. Rollo brought out the step-ladder, and placed it in the door-way, and then ascended it. From the top he could reach nearly to the top of the door; but then, as his tuft was at the end of the thread, it hung down, of course, some little distance below his head.

"Why, mother," said Rollo, "it goes *out*."

"Yes," repeated Nathan, "it goes out."

In fact, Rollo found that the tuft, instead of swinging into the room, was carried out towards the entry.

"You have found out, then," said his mother, "where the hot air of the room goes to, to make room for the cold air, that comes in from the entry."

"Yes, out into the entry," said Rollo.

"Through the upper part of the door," said his mother. "Suppose the entry were full of water, and the parlor full of air, and the door was shut, and the door and the walls were water-tight. Now, if you were to open the door, you see that the water, being heavier, would flow in, through the lower part of the door-way, into the parlor, and the air from the parlor would flow out, through the upper part of the door-way, into the entry. The water would settle down in the entry, until it was level in both rooms, and then the lower parts of both rooms would be filled with water, and the upper parts with air."

"Yes, mother," said Rollo.

"And it is just so with warm and cold air. If the parlor is filled with warm air, made so by the fire, and the entry with cold air, and you open the door, then the cold air, being heavier, will sink down, and spread over the floor of both rooms; and the warm air, being light, will spread around over the upper parts of both rooms; and this will make a current of air, in at the bottom of the door-way, and out at the top.

"Now," continued his mother, "let me recapitulate what I have taught you."

"What do you mean by *recapitulating* it?" said Nathan.

"Why, tell you the substance of it, so that you can write it down easier."

"O, I can write it now," said Rollo; "I remember it all."

"Can you remember it, Nathan?" said his mother.

"Perhaps I can remember some of it," said Nathan.

So Rollo and Nathan went out into another room, where Rollo kept his desk, and they remained there half an hour. When they returned, they brought their mother two papers.

Their mother opened the largest paper, and read as follows:—

"We took a tuft of down, tied to a thread, and held it in the cracks and places that the air came in at, to see which way it went. We held it at the window, and it blew *in* very strong. At the bottom of the door, it blew *in* very strong too; but at the top, it blew *out*, into the entry. So, when the entry is full of cold air, and this room full of warm, the cold air will press in and drive out some of the warm air, into the entry.

ROLLO."

The other paper was also in Rollo's handwriting, and was as follows:—

"If the entry was full of water, and the parlor full of air, and the walls were water-tight, and you were to open the door between the two rooms, the water would flow into the parlor down below, and the air would flow into the entry up above. We tried it with a tuft.

NATHAN."

QUESTIONS

Why were Rollo and Nathan at first glad to see the rain? What did their mother say to herself on the second day, when she observed their weary and listless appearance? What did she at first direct them to do? How did she prepare the downy tuft? What experiments did they perform with it? Where did they find that the air came in which crowded the warm air up the chimney? What experiments did they perform when the door was opened? Which way did they find that the current of air was setting at the lower part of the door-way? Which way did the current set at the upper part of the door-way? What did Rollo write in his exercise? What was written in Nathan's exercise?

CHAPTER IX
BURNING

After the snow had all gone off, and the ground was dry, Jonas piled up a heap of stumps, roots, and decayed logs, in a field, not far from the brook, and one sunny afternoon he and Rollo went down to set the heaps on fire.

Jonas set one on fire, and then he told Rollo that he might set another on fire. After this, Jonas employed himself in gathering up sticks, bushes, roots, and other such things that lay scattered about the field, and putting them upon the fires, while Rollo amused himself in any way he pleased.

After a time, Rollo found, on the margin of the field, near the edge of a wood, an old stump, taller than he was, much decayed. There was a hole in the top. Rollo climbed up so that he could put a stick in, and run it down, to see how far down the hole extended. He found that it extended down very near to the bottom.

Then Rollo called out to Jonas, with a loud voice, saying,—

"Jonas, I have found a hollow stump here. It is hollow away down to the bottom. May I build a fire in it?"

"Yes," said Jonas, "if you can."

Rollo accordingly went to the nearest fire, and got a quantity of birch bark, which he had collected there to aid him in kindling his fires. He lighted one piece, and put it upon the end of a stick, and carried it to the stump, with the rest of the birch bark in the other hand.

Rollo then spent some time in fruitless attempts to make some lighted birch bark go down into the stump, and burn there. He succeeded very well in getting pieces completely on fire; but, after they were dropped into the hole, they would not burn. Rollo could not think what the reason could be.

At last he called Jonas to come and help him set the stump on fire.

Jonas said that he did not think that it could be set on fire.

"'Jonas, I have found a hollow stump here,' said Rollo, calling with a loud voice"

"Why not?" said Rollo.

"Because," said Jonas, "it is so wet."

"Yes, but, Jonas," replied Rollo, "your brush heaps burn, and why should not this stump?"

"Because," said Jonas, "the stump is more solid, and the water soaks into it more in the winter and early in the spring; and it takes it much longer to dry, than it does brush and small roots, which lie open and exposed to the air."

"Well, then," replied Rollo, "why does not my birch bark burn? that is dry; but as soon as I drop it down into the stump, it goes out."

Jonas looked into the stump, and down around the bottom of it, and said,—

"Because there is no air."

"No air?" repeated Rollo.

"No," replied Jonas; "it is all close and solid around; the air cannot get in."

"It can get in at the top," said Rollo.

Jonas made no reply to this remark, but walked away a few steps, to a place where he had put down his axe; he took up the axe, and brought it to the stump. He immediately began to cut into it, at the bottom, as if it were a tree which he was going to fell.

"O Jonas," said Rollo, "don't cut it down."

"I am not going to cut it down," said Jonas; "I am only going to cut a hole into it."

"What for?" asked Rollo.

"To let the air in," replied Jonas.

Jonas continued to cut into the side of the stump, near the ground, until he perceived that the edge of his axe went through into the hollow part. Then he cleared away the chips a little, and showed Rollo that there was an opening for the air.

"Now," said he, "I presume you will be able to make sticks and birch bark burn in the stump, though you can't make the stump itself burn very well."

Rollo now dropped a blazing piece of birch bark into the stump, and, to his great joy, he found that it continued blazing, after it reached the bottom. He then dropped in another piece upon it, which took fire. He then gathered some dry sticks, and put in; and, finding that the flame was increasing, he proceeded to gather all the dry and combustible matter, which he could find around, and put them in, so that in a short time he had a fine blaze, a foot above the top of the stump; and the inside of the stump itself seemed to be in flames.

"Jonas," said Rollo, "it does burn."

"Does it?" said Jonas; "I am glad to hear it."

"But you said the stump would not burn."

"You ought to wait until it is all burnt up, before you triumph over me."

"Why, Jonas," said Rollo, "I didn't mean to triumph over you; but why would not the fire burn before you cut the hole through?"

"Because," replied Jonas, "there was not air enough."

"There was air in the stump," said Rollo.

"Yes," replied Jonas, "but all the life of it was consumed by the first piece of birch bark which you put in."

"The life of it?" said Rollo.

"Yes," replied Jonas; "what do you suppose it is, that makes anything burn?"

"Why, it burns itself," said Rollo.

"No," answered Jonas; "the air makes it burn: it must have good air around it, or else it won't burn. There is something in the air which I call the life of it; this makes the fire burn. But when this is all gone, then that air will not make fire burn any longer. It will only burn in good fresh air, which has got the life in it."

"I thought fire would burn in any kind of air," said Rollo.

"No," replied Jonas; "you can see if you stop up the hole I made here."

Jonas then took a piece of turf from the field, and put it before the hole, and crowded it in hard with the heel of his boot. Rollo observed that the fire was almost immediately deadened.

"Now," continued Jonas, "light a small piece of birch bark, and put it in."

Jonas helped Rollo fasten a small piece of bark upon the end of a stick, and then Rollo set it on fire, and held it down a little way into the stump. It burned very feebly.

"See," said Jonas, "how quick it is stifled."

"Yes," replied Rollo, "it goes out almost directly."

"You see," said Jonas, "that the fire already in the stump consumes all the goodness of the air; and I stopped up the hole, so that no fresh air can come in."

"Why doesn't it go in at the top?" said Rollo.

"It does a little," said Jonas, "but not much, because the hollow of the stump is already full of bad air, and there is nothing to make a current. When there is an opening below, then there is a current up through."

"Yes," said Rollo, "it is just like a chimney."

"Yes," replied Jonas, "the stump is the chimney, and the hole is the fireplace."

"And the air in the stump," said Rollo, "gets hot, and so the cold air all around is heavier, and so it crowds down under it, and buoys the hot air up out of the stump. My father explained it all to Nathan and me."

Rollo then wanted to open the hole again, to see if the effect would be as he had described.

Then Jonas pulled away the turf from the hole at the bottom of the stump, and Rollo observed that the fire brightened up immediately.

He then held a smoking brand near the hole, and he saw that the smoke was carried in, in a very strong current, by the cool air, which was pressing into the hole.

"Yes," said Rollo, "it operates just like a fireplace."

"So you see," continued Jonas, "that whenever you build a fire, you must see to it, that there is an opening for air to come up from underneath it. And it must be good fresh air too."

"What is it in the air, which makes the fire burn?" said Rollo.

"I don't know what the name of it is," said Jonas; "it is some part of the air, which goes into the fire, and is all consumed, and then the rest of the air is good for nothing."

"Isn't it good for anything at all?" asked Rollo.

"I don't know," said Jonas, "how that is; only I know that it isn't good for anything for fires. It stifles them."

"I should like to know what the name of that part of the air is, which is good for fires," said Rollo.

"I knew once," said Jonas, "but it was a hard word, and I have forgotten it."

"I mean to ask my father," said Rollo.

Jonas then went on with his work, gathering up everything that he could find around the field, to put upon the fires. Rollo amused himself by putting large rolls of birch bark around the end of a stick, and then, after setting them on fire, holding them over the fires, which Jonas was making, to see how soon the flame was extinguished: then he would draw them away, and see them revive and blaze up again in the open air. At last, he called out to Jonas, once more.

"Jonas," said he, "I have found out what makes the blaze go out. It is the smoke. I don't believe but that it is the smoke."

"No," replied Jonas, "it is not the smoke. I can prove that it is not."

So Jonas came up to the fire where Rollo was standing, and pointed out to Rollo a place, over a hot part of it, where there was no smoke, because the fire under it burned clear, being nearly reduced to coals. He told Rollo to hold his blazing bark there. Rollo did so, and found that it was extinguished at once, and as completely, as it had been before, when he had held it in a dense smoke.

"Yes," said Rollo, "it isn't the smoke. But perhaps it is because it is so hot."

"No," said Jonas, "it isn't that. It is a difference in the air. They sometimes collect different kinds of air in glass jars, and then let a candle down in, and see whether it will go out."

"And will it go out?" said Rollo.

"That depends upon what kind of air it is," said Jonas. "They all look clear, just as if there was nothing in the jars; but when you let a candle down in, in some it burns just the same as before; in some it burns brighter; and in some it goes out."

"In what kinds does it go out?" asked Rollo.

"I only know of one kind," said Jonas, "and that is a kind that comes of itself in mines, and wells, and other places."

"What is the name of it?" asked Rollo.

"Why, the people in the mines call it choke damp; but I believe it has got another name besides."

"What do they call it choke damp for?" said Rollo.

"Because," said Jonas, "if the miners get into it and breathe it, it kills them. It is not any better to breathe than it is to make fires burn."

"I wish I could see some choke damp," said Rollo.

"O, you can't see it at all," said Jonas, "if it was right before you, any more than you can see common air. If a well or a mine is full of it, they cannot find it out by looking down."

"How do they find it out?" said Rollo.

"Why, they let a candle down," replied Jonas.

"And will the candle go out?" asked Rollo.

"Yes," said Jonas, "if there is choke damp in the well. Sometimes they make a little of it in a tumbler or a jar upon the table, and so let a little flame down into it, and it goes out immediately."

"I wish we could make some," said Rollo. "Do you know how they make it?"

"No," said Jonas; "but I believe it is pretty easy to do it, if we only knew how."

"I will ask my father," said Rollo; "perhaps he will know."

This conversation took place when Jonas and Rollo were about the fires; but now the fires had pretty nearly burnt out, and they prepared to go home.

That evening, just about sunset, Rollo went out behind the house, and found Jonas raking off the yard. The spring was fast coming on, and the grass was beginning to look a little green; and Jonas said he wanted to get off all the sticks, chips, and straws, so that the yard would present a surface of smooth and uniform green. Rollo told him that he had found out how to make choke damp.

"Did your father tell you?" said Jonas.

"No," replied Rollo.

"Who did tell you, then?" said Jonas.

"Guess," answered Rollo.

"Your mother," said Jonas.

"No," answered Rollo.

"Then I can't tell," said Jonas.

"It was Miss Mary," replied Rollo. "I met her in the road to-day, and I asked her."

"And how is it?" asked Jonas.

"Why, we make it with chalk and vinegar," said Rollo. "We pound up a little chalk, and put it in the bottom of a tumbler. Then we pour some vinegar over it. The vinegar takes the choke damp out of the chalk, and Miss Mary says it will come up in little bubbles. She says we can lay a paper over the top loosely,—she said loosely, but I think it ought to be tight."

"Why?" asked Jonas.

"So as to keep the choke damp from coming out," replied Rollo.

"No," said Jonas. "I understand why she said you must put it on loosely; that's to let the common air out."

"What common air?" said Rollo.

"Why, the air that was in the tumbler before," replied Jonas. "You see that, as fast as the choke damp comes up, it drives the common air out of the top of the tumbler; and so you must put the paper on loosely, and let it go out."

That evening Jonas and Rollo tried the experiment. First they put about two teaspoonfuls of chalk into the tumbler. Then they poured in the vinegar. It immediately began to foam.

"Ah," said Rollo, "that's the effervescence."

"The what?" said Dorothy; for they were making this experiment upon the kitchen table, and Dorothy was standing by, looking on with great interest.

"The *effervescence*," said Rollo. "Miss Mary said there would be an effervescence, which would be occasioned by the little bubbles of choke damp, coming up from the chalk."

"Poh!" said Dorothy; "it's nothing but a little frothing."

"It isn't frothing," said Rollo, very seriously; "it isn't frothing, it is effervescence. Don't you think Miss Mary knows?"

"Jonas," said Rollo again after a short pause, "how many of these little bubbles will it take, do you think, to fill the tumbler full of choke damp?"

"I don't know," replied Jonas; "we will wait a little while, and then try it."

"There, now, Jonas," said Rollo, "we have not got any candle."

"O, I will roll up a piece of paper, and set the end on fire, and then dip it down into the tumbler, and that will do just as well."

"What are you going to do that for?" said Dorothy.

"Why, to see it go out," said Rollo.

"It won't go out, unless you put it away down into the vinegar," said Dorothy.

"Yes it will," said Rollo; "we are only going to dip it down a little way, just into the choke damp, and it will go out."

"It won't go out, child," said Dorothy. "There's nothing to put it out."

"Well, you'll see. Won't it go out, Jonas?"

"I don't know," said Jonas.

"Don't know?" said Rollo. "Why, you told me that choke damp would put out a blaze."

"Yes," said Jonas, "I am sure of that; but there are a great many ways of failing in trying experiments."

"Well," said Rollo, "that may be; but this will not fail, I know, for I can see the little bubbles of choke damp coming up. There are millions of them."

By this time Jonas thought that the tumbler was filled with the gas, which was rising from the chalk and vinegar. So he rolled up a piece of paper, and set the end on fire, and, when it was well burning, he plunged the end of it into the tumbler. To Rollo's great disappointment and mortification, it continued to burn about as much as ever. The flame crept rapidly up the paper, and Jonas had soon to run with it across the floor and throw it into the fire, to avoid burning his fingers. Dorothy laughed aloud; Jonas smiled; and as for Rollo, he looked disappointed and vexed, and appeared to be overwhelmed with chagrin.

Dorothy continued to laugh at them, while Jonas went to the pump and washed out the tumbler. At length she said,—

"But come, Rollo, don't be so disconsolate. You look as if you had swallowed all the choke damp."

"Yes, Rollo," said Jonas, "we must keep good-natured, even if our experiments do fail."

"Well," said Rollo, "I mean to ask Miss Mary again, and then we can do it, I know."

Rollo accordingly went, the next day, to ask Miss Mary about the cause of the failure. Miss Mary said that she could not think of any thing which was likely to be the cause, unless it was that they put too large a flame into the tumbler.

"Well," replied Rollo, "and what harm would that do? Won't the choke damp put out a large flame?"

"Yes," replied Miss Mary, "if it only fairly surrounds and covers it; but, then, if you put a large flame into a tumbler, it makes the first instant, a great current of air, and so the choke damp might be blown out, and common air get in, and so keep the paper burning."

"How does it make a current of air?" asked Rollo.

"Why, the heat of the flame, when you first put the paper in," replied Miss Mary, "makes the air that is above it lighter; and the common air all around crowds in under it, in buoying it up; and by that means, if the flame is too large, common air is carried into the tumbler. You ought to make a very small flame, if you leave the top of the tumbler open."

"How can we make a small flame?" said Rollo.

"One good way," replied Miss Mary, "is to roll up some paper into a very small roll. I will show you how."

So Miss Mary took a piece of paper, and cut it into the proper shape with her scissors, and then rolled it up into a long and very slender roll; one end of it was not much larger than a large knitting-needle. She gave this to Rollo, and told him that, if he tried the experiment again, he must light the small end, and it would make a flame not so big as a pea.

Rollo explained to Jonas what Miss Mary had said, and they resolved on attempting the experiment again that evening. And they did so. Dorothy stood by watching the process, as she had done the evening before, but Rollo did not assert so confidently and positively what the result would be. He had learned moderation by the experience of the night before.

When all was ready, Jonas lighted the end of the slender roll in the lamp, and plunged it carefully into the tumbler. It went out immediately.

"There!" said Rollo, clapping his hands, "it goes out."

"Why, it is only because the wind blew it out."

"No, Dorothy," said Rollo, "there isn't any wind in the tumbler."

"Yes," replied Dorothy, "when you push it down, it makes a little wind, just enough to blow it out."

"Get another tumbler," said Jonas, "and let us see."

So Dorothy brought another tumbler, and Jonas put the burning end of the paper down into it, with about as rapid a motion as that with which he had put it before into the tumbler he had at first. The paper continued to burn.

"There," said he to Dorothy, "when I put it down into common air, it burns on the same as ever; so it can't be that the wind puts it out." Jonas repeated the experiment a number of times; the effect was always the same. Whenever he put it into the tumbler of common air, it burned on without any change; but whenever he put it into the choke damp, it immediately went out. Even Dorothy was satisfied that there was a difference in the kind of air contained in the two tumblers.

That evening, when Rollo gave his mother a full account of their attempts,—describing particularly their failure at first, and their subsequent successes,—his mother seemed much interested. When he had finished, she said,—

"Well, Rollo, I don't see but that you have learned two lessons in philosophy."

"Two lessons?" said Rollo.

"Yes," replied his mother. "The first is, that fire will not burn in choke damp; and the second is, that it requires nice attention and care to verify philosophical truths by experiment."

"Yes," said Rollo, "we missed the first time, just because we had too big a paper."

QUESTIONS

Why did Jonas suppose that the stump would not burn? What was Rollo's first mode of setting it on fire? How did it succeed? What did Jonas do with his axe, when he came? What was the object of this? What did he say was necessary to make fires burn? What did Rollo at first think was the reason why the bark went out when held over the fire? What did he next think was the reason? How did Jonas say that different kinds of airs were prepared? In what places did he say that choke damp was naturally produced? How did they attempt to prepare some of this gas? Did they succeed in preparing it? Did they succeed in their experiment at first? What was the cause of the failure?

CHAPTER X
GRAVITATION

One evening, after tea, when Rollo was a pretty big boy, he came and began to climb up into his father's lap. When he had climbed up, he took his place astride of his father's knee, as if he were riding a horse. His little brother Nathan came up and stood near, wanting to get up too, only there was not room. His cousin James was there, that evening, on a visit. He sat upon a cricket before the fire, and his mother was at the table doing some sort of work.

"O dear me!" said Rollo's father, imitating the tone in which Rollo sometimes uttered that exclamation.

"What, sir?" said Rollo.

"Why, I should like very well to hold you in my lap," said his father, "if it was not for the great mighty earth, down below us."

"How?" said Rollo. He did not know what his father meant.

"Why, when you are upon my knee, the earth, the ponderous earth, pulls you down hard and heavy upon it." So saying, he put his hands upon Rollo's shoulders, and crowded them down, by way of showing him how the earth acted upon him. "It pulls," he continued, "with a strong and steady pull, all the time; and so makes you a very heavy weight."

"Is that what makes weight?" said Rollo.

"Yes," said his father. "So, if I had a monstrous stone to move, and if I thought the earth would listen to me, and let go its hold, I might make a speech to it thus:—

"'O earth, thou vast and ponderous ball, please to relax thy hold, for a few minutes, upon this stone, and leave it free to move; and then Rollo can tie a string to it, and move it easily along to the place where I want it to lie; then thou mayst seize it again with thy mighty attraction, and hold it down as firmly as thou wilt.'"

"O father!" exclaimed Rollo; Nathan and James laughed, and Rollo's mother looked up from her work to listen to this strange apostrophe.

"It would seem," continued his father, in a pompous tone, as if still addressing the earth—"it would seem, most mighty planet, a very easy thing for thee to release this single stone, for a few minutes, from the grasp with which thou holdest all things down upon thy surface. And by it I shall gain much, while thou wilt lose nothing; for, if thou wilt not willingly give up the stone, I must get three or four yoke of strong oxen, and, by main force, pull it away."

"Is that what makes everything heavy?" said Rollo.

"Yes," said his father, answering now in his natural tone; "the attraction of the earth is what makes everything heavy, and holds it down."

"And could we move a monstrous great stone," said Rollo, "as light as a feather?"

"No," said his father, "it would not move along quick and light, like a feather. You could not move it quick. Suppose, for instance, you had two boats, floating upon the water, of the same size; one made very light indeed, of something very thin, like paper, and empty; and the other made of wood, and loaded with iron as heavily as it would bear. Now, they would both be supported upon the water, so that their weight would be neutralized; and yet they would move very differently. You could push the light one about easily, anywhere, but the heavy one would move very slowly. You would not have to push very hard upon it, but you would have to push *for some time*, to set it in motion; and then it would be hard to stop it. This is called its *inertia*."

"Yes," said Rollo, "it would go harder against the bank."

"The reason is," continued his father, "that the heavy boat contains a great many more particles of matter than the light one, and they have all got to be put in motion. So it requires greater effort, or the same effort must be continued a longer time.

"For instance, if we suppose that the light boat has one million of particles of matter, the heavy one would have, perhaps, twenty millions. Of course the effect of the pushing has to be divided among twenty times as many particles, and of course will only carry them one twentieth part as far; so that the bodies that are now large and heavy, would only move slowly, though they would move *easily*, if the attraction of the earth were to cease.

"There is another way to illustrate it," he continued. "Suppose there was a large mass of lead, as big as a load of hay, hanging by a chain; and also a great puff of feathers, or a balloon of the same size, hanging in the same way. Now, if they were both suspended freely, they would both move easily, for their weight would be supported by the chain; but the heavy one would move very slowly. Nathan could move it, but he could only move it slowly and a little way."

"I should not think that he could move it but very little," said Rollo.

"No, he could not; because you see that, in that way of suspending anything, the moment that it begins to move, it begins to swing off and to rise; so that it cannot be moved at all without being *lifted* a little. And the more it is moved, the higher it is lifted, so that it would take a great force to move it far away from the centre, where it was hanging. But we can hang it in a way to avoid that difficulty."

"How, sir?" said Rollo.

Rollo seemed to be very much interested in this conversation. He had dismounted from his father's knee, and stood by his side, listening eagerly. His mother, too, was paying close attention. As for Nathan, he sat still; though it is not by any means certain that he understood it very well.

"Let us suppose," said his father, "that the mass of lead, as big as a load of hay, is fastened to one end of a stick of timber."

"That would not be strong enough to hold it," said Rollo.

"Well, then, to a beam of iron, as large as a stick of timber," rejoined his father.

"O," said James, "you could not get such a big bar of iron."

"No," replied his father, "only an imaginary one; and that will be just as good as any. Now, suppose the great mass of lead is fastened to one end of this bar, and another one, just like it, to the other end, to balance it. Now, suppose that the lower end of the great chain is secured around the middle of the iron beam, and the upper end to be fastened to some strong support up in the air. Now, we can move the mass of lead without having to lift it at all; for, if we push against it, and make it move, it will move round and round, without rising at all, as it did before, when it was hung up directly by the chain."

Rollo's father then went on to explain to them that, in such a case as this, the weight of the two masses of lead would not prevent their moving easily, for they would exactly balance each other. A little child would be able to move them; but still they would move exceedingly slow at first, and it would be hard to stop them, when they were in motion. So, he said, if the earth should cease to attract and draw down any great, heavy body, like a large stone, for example, the smallest child could lift it, though it would come up slowly, just as a very heavy body would move, if it was suspended by a string, or was afloat upon the water.

"And so," said Rollo, "if the earth should not attract us, could we push ourselves right up off from the ground?"

"Yes," said his father, "most undoubtedly."

"What, and go about anywhere in the air?"

"Certainly."

Rollo began to laugh aloud at this idea, and looked very much interested and pleased.

"O, then I wish there was no gravitation," said Rollo; "I do, really."

"But, then," continued his father, "if you should get up into the air, you could not get down again."

"Why not?" said Nathan, beginning to look a little concerned.

"Unless," said his father, "you had something above you, to push against, so as to push yourselves down. You would be just like a boy in a boat, off from the shore, and without any paddle or pole. He could not get back again."

"We might tie a rope to something," said James, "before we went up, and so pull ourselves down."

"Yes, that you might do."

"And could not we flap our hands, like a bird, and so fly a little?"

"Perhaps you could," said his father.

Here the children all began to flap their hands, like young birds trying to fly; and Rollo said again, he wished, with all his heart, there was no gravitation; "for then," said he, "we should have strength enough to fly."

"That would lead to serious consequences," said his father.

"What consequences?" said James.

"Much more serious than you would suppose."

"Tell us what they would be, uncle," said James.

"O, I know," said Rollo; "you could not stand up straight without gravitation."

"O, we could, couldn't we, father?" said Nathan.

"What makes you think, Rollo," said his father, without replying to James's question—"what makes you think that we could not stand up straight without gravitation?"

"Why, you see," said Rollo.—Here he paused, and looked confused, and did not know what to say. He had an indistinct recollection of having read something about it in some book; but he could not tell what.

"I don't see what should prevent any body's standing up straight, if the attraction of the earth should cease; in fact, if it made any difference, it would be rather easier to stand up straight."

Here Rollo looked rather foolish, but he did not reply. The truth is, like almost all other children, who take an interest in reading, he was sometimes a little vain of his knowledge; and in this case, instead of listening attentively, and endeavoring to learn something new from his father's explanations, he seems to have thought it a good plan for him to help him elucidate the subject to James and Nathan. He exchanged the character of learner for teacher too soon.

"Well, uncle," said James, "what would be the consequence if gravitation should cease?"

"Why, in the first place," said Rollo's father, "all the streams in the world would stop running."

"The streams!" said Rollo, astonished.

"Yes," said his father, "every river, brook, and rill. The reason why the streams flow is, that the earth attracts the water from the mountains and hills, down into the valleys and towards the sea."

"Well, sir, what else?" said Rollo.

"Why, there would never be any more rain."

"No more rain!" exclaimed all the children.

"No," he replied. "The drops of rain fall only because the earth draws them down by its attraction; and, of course, if this attraction should cease, they would remain where they are."

The children were musing a minute upon these strange effects, when Rollo asked if anything else would happen.

"Why, yes," said his father, "worse disasters than these; but I do not know whether you would understand them, if I should explain them."

"O, try," said Rollo; "I think we shall understand."

"Well, let me think," said his father. "You have noticed how a chaise wheel, on a muddy road, in a wet day, holds the mud upon it, until when it is going very swiftly down a hill, and then the mud flies off in all directions."

"Yes, sir," said all the children.

"And if the mud did not stick to the wheel pretty tight, it would be thrown off at all times, even when the wheel was going slow. You understand this."

"Yes, sir."

"Well, now, this whole earth, you all know, is whirling around through space, and moving on also around the sun. And all the loose things upon the surface would be thrown off at once, if they were not held to it by a strong attraction. If this attraction were to cease suddenly,—whisk!—away we should all go in an instant—rocks, houses, men, animals, all in confusion."

"O father!" exclaimed Rollo; "where should we go to,—off into the air?"

"Not exactly into the air, for the air would all fly off, and be dissipated too; we should fly off into the sky somewhere, some in one direction, and some in another. You'd be a thousand miles off from the earth, almost before you would know it."

"Would it kill us, father?" said Nathan.

"Yes," said his father. "I don't know that there would be any shock that would hurt us, but we should have no air to breathe, and it would be dark and dismal."

"Dark?" said Rollo. "There would be the sun."

"Yes," said his father, "there would be the sun; and the sun would look bright enough when you looked directly towards it, but there would be no general light about you, unless there was air."

The children all paused to reflect upon the strange results which their father had told them would ensue from a suspension of the earth's attractive force. Rollo began to think that he had been too hasty in his wish that there was no gravitation.

"But, father," said he, "the houses would not go off, certainly;—only the loose things would go."

"Very well; houses are loose."

"O father! they are fastened down."

"How are they fastened down?" asked his father.

"O, they are nailed—and,"

"Not nailed to the ground, certainly," said his father.

"No," said Rollo, laughing; "but then they are built with great stones and mortar."

"Yes, but there is no mortar under the lowest stones. The foundations are simply laid upon the ground."

"Well," rejoined Rollo, "I thought they were fastened somehow or other."

"No," said his father; "they dig the cellar, and only just lay the foundations upon the ground, without any fastening. The earth holds them in place."

"Well, father," said Rollo, "that is what I meant, when I said we could not stand up straight. I meant the houses. I read in a book that houses would be blown away, if the gravitation did not hold them down."

Here Rollo's father had a hearty laugh; and he told Rollo that he thought that was rather wide shooting. Rollo wanted to know what he was laughing at; and Nathan asked him what he meant by wide shooting.

"Why," said he, "Rollo, you undertook to explain to us, from your stores of knowledge, what the effects of a suspension of gravitation would be; and, in attempting to tell that houses would be in danger of being blown away, you came no nearer than to tell us that boys could not stand up straight; and that is what I call pretty wide shooting."

So saying, he rose from his seat, and walked away, appearing to be very much amused. James laughed too, and even Rollo could not help smiling at the ridiculous figure which his display of his learning made. As for Nathan, he continued to look grave; and said he did not see that it was any shooting at all.

After a short pause, Rollo's mother said, "So you see, children, the cause of all the pressure, both of air and of water, and all the effects produced by them, are the results of their gravitation towards the earth."

"Yes," said Rollo, "I believe I understand it now."

After this, Rollo took James and Nathan out into the yard, to see if some beans had come up, which he had been planting in a sunny corner of the garden the day before.

QUESTIONS

What was Mr. Holiday's apostrophe to the earth? What is the cause of weight? Why did the boys wish that there was no gravitation? What was the first evil consequence which their father said would ensue, if there was no gravitation? What was the second evil consequence? What did their mother say after the conversation with their father was closed?

CHAPTER XI
AIR IN MOTION

Rollo's dam, which he had made when he was studying the philosophy of water, and which was at first undermined by the pressure of the water, was afterwards carried away by its momentum. Rollo learned, at that time, that water moving rapidly had a great momentum; and about this time he had an opportunity of learning that air, when in motion, had a momentum too, capable of producing very powerful effects. The circumstances of the case were as follows:—

One morning, towards the latter part of March, Jonas, being out in the barn, observed some indications that the roof wanted repairs. It had been strained and weakened by the heavy snows in the winter. He reported the fact to Rollo's father, who said that he might go, the next day, and get the carpenter to come and repair it. The carpenter lived ten miles distant, near the shore of a long pond.

When Rollo heard of this proposed expedition, he wanted to go too; and his father gave him permission. Jonas was going in the wagon. He told Rollo, the evening before, that he meant to set out at six o'clock.

"But suppose it looks like a storm," said Rollo.

"Then there will be more need of going," said Jonas; "for if the equinoctial storm comes on before the roof is strengthened, it may get carried away."

"What is the equinoctial storm?" said Rollo.

"O, it is a great storm, which comes generally about this time of year. I shouldn't wonder if it should come on to-morrow. But it may not come for a week; and so I hope we shall have time to get the roof mended first."

"Does it look like a storm to-night?" said Rollo.

"No, not much," replied Jonas. "It is a little hazy in the south-west. However, if it looks like a storm in the morning, you need not go, unless you choose; though I shall."

"I wish you'd wait till the storm is over," said Rollo.

"No," said Jonas, "I had rather go in the storm than not."

"Why?" said Rollo.

"Because," said Jonas, "I like to be out in storms. Sometimes it is very grand."

The next morning, when Rollo awoke, he found that it was light, but not yet sunrise. He arose, and looked out of the window to see if it was pleasant. The sky was somewhat overcast, but there was a little blue to be seen, and Rollo thought that it would be pleasant. He heard a noise in the barn-yard, and, looking in that direction, he saw Jonas just leading the horse out of the stable. So he dressed himself soon, and went down.

When he got ready, he went down into the yard, and found that Jonas had got the horse harnessed, and everything prepared. There was a little bag of oats in the back part of the wagon, and also a tin pail, with a cover, which contained a luncheon. Jonas fastened the horse to a post, and said,—

"Now, Rollo, we'll go in and get some breakfast."

"I thought that luncheon was for breakfast," said Rollo.

"No," said Jonas, "that is for dinner."

"Shall we be gone all the day?" said Rollo.

"We may be gone till after dinner," said Jonas, "and so I thought I would be sure."

The two boys went into the house, and there they found that Dorothy had got some breakfast ready for them upon the kitchen table. After eating their breakfast, they got into the wagon, and set out. Jonas first put in a large umbrella. Just as they were driving out of the yard, the first beams of the morning sun shone in under the branches of a great tree in the yard, and brightened up the tips of the horses' ears and the boys' faces. At the same time, a rude gust of wind came around the corners of the house, and slammed to the gate of the front yard.

"It's going to be pleasant," said Rollo; "the sun is coming out."

"I'm not very sure of that," said Jonas; "the wind is rising."

"We start just at sunrise," said Rollo.

"Yes," replied Jonas, "the sun always rises at six o'clock at this time of the year."

The boys rode along for about three hours, before they came to the carpenter's. They were obliged to travel very slow, for the roads were not good. It is true that the snow was all gone, and the frost was nearly out of the ground; but there were many deep ruts, and in some places it was muddy. The sun went into a cloud soon after they set out, and it continued overcast all the morning. There was some wind too, but, as it was behind them, and as the road lay through woods and among sheltered hills, they did not observe it much. Jonas said that there was a storm coming on, but he thought it was coming slowly.

They arrived at length at the pond. There was a little village there, upon the shore of the pond. The reason why there happened to be a village there, was this: A stream of water, which came down from among the mountains, emptied into the pond here, and, very near where it emptied, it fell over a ledge of rocks, making a waterfall, where the people had built some mills. Now, where there are mills, there must generally be a blacksmith's shop, to mend the iron work when it gets broken, and to repair tools. There is often a tavern, also, for the people who come to the mills; and then there is generally a store or two; for wherever people have to come together, for any business, there is a good place to open a store, to sell them what they want to buy. Thus there was a little village about these mills, which was generally called the Mill village.

Jonas inquired where the carpenter lived, and then drove directly to his house. He found that he was not at home. He had gone across the pond, to mend a bridge, which had been in part carried away by the floods made when the snow went off. Rollo sat in the wagon in the yard by the side of the carpenter's house, while Jonas stood at the door, making inquiries and getting this information.

"If you want to see him very much," said the carpenter's wife, "I presume you can get a boat down in the village, and go across the pond."

"How far is he from the other side of the pond?"

"O, close by the upper landing," said she; "not a quarter of a mile from the shore, right up the road."

Jonas thanked the woman for her information, and got into the wagon.

"Let us get a boat and go over, Jonas," said Rollo, as they were turning the wagon round.

"I should," said Jonas, "if there was not such a threatening of a storm."

"It does not blow much," said Rollo.

"No," said Jonas, "not much now, but the wind may rise before we get back. However, we'll go and see if we can get a boat."

"And very soon they were gliding smoothly along out of the cove"—

After some inquiry, they found a boat, at a little distance out of the village, in a sort of cove, where there was a fine, sandy beach. The boat was of very good size, and it had in it two oars and a paddle. Jonas looked out upon the water, and up to the sky, and he listened to hear the moaning of the wind upon the tops of the trees. He wanted very much to persevere in his effort to find the carpenter; but then, on the other hand, he was not sure that it was quite safe to take Rollo out upon the water at such a time. He sat upon a log upon the shore a few minutes, and seemed lost in thought.

At last he said,—

"Well, Rollo, I believe we'll go. The worst that will happen will be, that you may get frightened a little. We can't get hurt."

"Why can't we get hurt?" said Rollo.

"Why, even if it comes on to blow hard, it will probably be a steady gale, and I can run before it, if I can't do anything else. And there can't be much of a sea in this pond."

Rollo did not know what Jonas meant by much of a sea in the pond; but, as Jonas immediately went to work taking the horse out of the wagon, Rollo did not ask any questions. The boys unharnessed the horse, for Jonas said he would stand easier out of harness, and they might be gone more than an hour. They fastened him then to a tree, and poured the oats down before him upon the ground. Then Jonas helped Rollo into the boat, and put in the tin pail containing their luncheon, and also the umbrella; though he said he did not think it would rain before they got back. Then he shoved off the boat, and jumped in himself; and very soon they were gliding smoothly along out of the cove.

Rollo wanted to row; and so Jonas let him take one oar, while he himself sat in the stern with the paddle. Rollo soon learned the proper motion, so that his efforts assisted considerably in propelling the boat. They found, when they were out at a little distance upon the water, that the wind blew much harder than Rollo had expected.

"Jonas," said he, "the wind blows more here than it did upon the shore."

"No," said Jonas, "only we feel it more here than when we were under the lee of the land."

"What do you mean by the lee of the land?" said Rollo.

"I mean the shelter of it," replied Jonas. "Whenever a ship at sea is sheltered by anything, they say the ship is under its lee."

The boys went on, Rollo rowing, and Jonas paddling behind, until at length Rollo got tired. Jonas then told him to spread the umbrella, and hold it up for a sail. Rollo did so. The wind was blowing pretty nearly in the direction in which they were going, and, by its impulse upon the umbrella, it caused it to pull very hard. Rollo rested the middle of the handle of the umbrella upon his shoulder, holding the crook in his hand, turning it in such a position as to present the open part of the umbrella fairly to the wind. Jonas continued to paddle, and so they went on very prosperously until they had got two thirds across the pond, when Jonas ordered Rollo to take in sail.

"Why," said Rollo, "we have not got across yet."

"No," replied Jonas, "but the wind is taking us out of our course."

Rollo drew down the umbrella, and looked around. They were still at a considerable distance from the shore. Jonas extended his paddle out into the water as far as he could reach, and then drew it in towards him with several quick and strong strokes, as if he were endeavoring to pull the stern of the boat, in which he was sitting, round.

"What are you doing so for?" said Rollo.

"I am trying to bring her up into the wind," replied Jonas.

"What is that for?" asked Rollo.

"Why, we've drifted to leeward," said Jonas, "and I must bring her *up*; for we want to land around behind that point on the starboard bow."

Rollo did not understand Jonas's technical language very well. He particularly did not know what Jonas meant by bringing her *up*, for it seemed to him that the pond was perfectly level, so that there was no up or down either way. He did not know that, in sea language, *against* the wind was always *up*, and *with* the wind, *down*.

Jonas found it hard to bring the boat up into the wind. The waves had begun to be pretty large, and they beat against the bows of the boat, and some of the water dashed over upon Rollo. The wind blew quite heavily, too; and now that they had changed their direction so as to bring the wind upon their side, it embarrassed, if it did not absolutely retard their progress. Some drops of rain also began to fall.

However, by hard and persevering exertion, Jonas at length succeeded in urging the boat forward until he began to draw nigh to the point of land; and soon afterwards they came under the shelter of it, where the water was smooth, and the air comparatively still. Here Rollo put in his oar again, and they passed along close under a high shore, for some distance, until they came to the landing. Here they fastened the boat, and then began to walk along up the road.

The road lay through the woods, and among hills, so that it was sheltered; and the only indications of the wind which the boys noticed, was a distant roaring sound among the forests. They came at length to the bridge, where they found several workmen busily engaged in laying abutments of stone, but the carpenter himself was not there. The men told Jonas that he had gone about half a mile away, on a by-road, to select and cut some timber to be used in the construction of the bridge.

"How long will he be gone?" asked Jonas.

"He will be gone two or three hours," said a man with a stone hammer in his hand.

"What shall we do now?" asked Rollo, addressing Jonas, after a short pause.

"Keep on until we find him," replied Jonas. "But you may stay here and see them build the bridge, while I go after the carpenter."

Accordingly Jonas went on, leaving Rollo seated upon a bank watching the work. In about three quarters of an hour, he returned; and then he and Rollo went back to the boat. The wind had all this time continued to increase, though they were so much sheltered, that they did not notice it much.

Jonas, however, observed that some light, scudding clouds were flying across the sky, very low, being apparently far beneath the other clouds. When they reached the boat, Rollo proposed that they should stop and eat some luncheon; but Jonas said that he should eat his with a better appetite on the other side of the pond. So he hastened Rollo into the boat, and, talking his station in the stern, he began to ply his paddle with all his force, running the boat along under the shelter of the high shore.

"There isn't much wind, Jonas," said Rollo.

"We can tell better when we come round the point," replied Jonas.

Rollo observed that Jonas looked a little anxious, and he also seemed to be exerting himself so much in the long, steady strokes of his paddle, that it appeared to be rather an interruption to him to hear and answer questions. Rollo therefore did not talk. He found, however, as he drew near the point, that the waves were running by it, with great speed and force, down the pond. As the boat shot out from the shelter of the point into this place of exposure, the storm struck them suddenly, with a blast which swept the bows of the boat at once round out of her course, and dashed the spray from the waves all over Rollo's face and shoulders. It was with great difficulty that Jonas could bring the boat to the wind again.

He succeeded, however, at length, and they went on, for some time, pitching and tossing, through the waves,—the wind pressing so hard upon the boat that it was very difficult for Jonas to make any headway. The wind had changed its direction, so that it blew now almost exactly across their course; and it required great exertion for Jonas to prevent being blown away down the pond, out of his track altogether.

In the mean time, the wind rather increased than diminished; and the water dashed in so much over the bows that Rollo had to dip it up with the cover of the tin pail, and pour it out over the side of the boat into the pond again. They were going on in this way, both toiling very laboriously, when suddenly they began to hear a sound like distant thunder, somewhat louder than the ordinary roaring of the wind. They both looked towards the shore in the direction from which the sound came. On the declivity of a range of hills covered with forests they saw an unusual commotion among the trees. The tops were bowed down with great force; the branches were broken off, and Jonas thought that he could see fragments of them flying in the air; and presently, farther down, he observed several tall pines bending over, and then sinking down till they disappeared.

"What is it?" said Rollo.

"A squall," said Jonas,—"and coming down directly upon us."

"What shall we do?" asked Rollo.

"Put the boat before the wind," replied Jonas, "and let her run: we must go where the squall carries us."

Jonas immediately began to pull the stern of the boat around with his paddle, so as to turn the head of it away from the quarter which the wind was blowing from; and then the wind drove the boat along very rapidly over the waves, which curled and foamed on each side, driving onward with great fury. When they looked around behind them, they saw that the pond, which was of a very dark color, though spotted with the white tips of the waves all over its surface, was almost black for a large space in the direction from which the squall was coming. It advanced with great rapidity, and at last struck the boat with a noise like thunder. The froth and foam flew over the surface of the water like tufts of cotton, and the boat seemed to fly along the water with almost as much speed as they; and the roaring of the winds and waves was so loud that Rollo had great difficulty in making Jonas hear what he had to say. After a few minutes, the violence of the wind somewhat abated; but it still blew a steady and furious gale, so that Jonas had to keep his boat directly before it. Thus they were driven on, wherever the wind chose to carry them, for more than half an hour.

Then they began to draw near the land, far, however, very far from the place where they had intended to go. Rollo observed that Jonas was looking out very eagerly towards the shore, and he asked him what he was looking for.

"Why, here we are," said Jonas, "on a lee shore, and I am looking out for a place to land."

Rollo looked, and saw that the waves were tumbling with great violence upon the rocks and gravelly beaches which lined the shore, and he was afraid that the boat would get dashed to pieces upon them. Jonas, however, observed a large tree, which originally stood upon the bank, but which had fallen over, and now lay with its top partly submerged. He thought that this might afford him some shelter, and so he made great exertions to guide the boat so as to bring it in to the shore around behind this tree. By means of great efforts he succeeded; and so he and Rollo both escaped safe to land.

The boys did not get home until late that night, for they were thrown upon the shore nearly two miles from the Mill village, and of course they had that distance to walk. Jonas was detained a little there, too, in making arrangements to send a boy for the boat after the storm had subsided. When they got home, Rollo's father said that he was sorry for their fatigues and exposures, but he was very glad that Jonas had persevered and found the carpenter; for the high wind had blown down the back chimney and broken the roof over the kitchen, and it was very necessary to have it repaired immediately.

QUESTIONS

What is *momentum*? Has air momentum, when it is in motion, as well as water? At what time in the spring of the year does the sun rise at six o'clock? What did Rollo think was the prospect in respect to the weather? What did Jonas think? What is meant by being under the *lee* of a shore? What is a *squall*? What indications did Jonas observe of the approach of the squall? What course did he pursue in order to avoid the danger of it?

CHAPTER XII
AIR AT REST

A few days after the adventure described in the preceding chapter, Rollo heard his father proposing to his mother that they should take a walk the next morning before breakfast. Rollo wanted to go too. His father said that they should be very glad to have his company; and he promised to wake him in season.

Rollo felt rather sleepy, when his father called him the next morning; but he jumped up and dressed himself, and was ready first of all. It was a cool, but a very pleasant morning. The sun was just coming up. The ground in the path before the door was frozen a little, and the air seemed very still.

When Rollo's mother came out to the door, she said,—

"Well, husband, which way shall we go?"

"Up on the rocks," said Rollo; "let's go up on the rocks, mother. It will be beautiful there this morning."

"Well," replied his mother; "we'll go up on the rocks."

The place which Rollo called the rocks, was the summit of a rocky hill, which had a grassy slope upon one side, by which they could ascend, and a precipice of ragged rocks upon the other. There was a very pleasant prospect from the top of the rocks.

As they walked along, Rollo said that it was very different weather that still morning, from what it was the day that he and Jonas were out upon the pond.

"Yes," said his father, "you had an opportunity to see the effects of air in motion then."

"And now *air at rest*," replied Rollo.

"Pretty nearly," said his father.

"Yes, sir, *entirely*," said Rollo; "there is no wind at all, this morning: hold up your hand, and you can feel."

So Rollo stopped a moment upon the grass, and held up his hand to see whether there was any wind.

"I know there is not any wind that you can perceive in that way," said his father.

"How can we perceive it, then?" said Rollo.

"I'll tell you," replied his father, "when we get to the top of the hill."

They reached the top of the hill soon after this, and sat down upon a smooth stone. There was a very wide prospect spread out before them,—fields, forests, hamlets, streams,—and here and there, scattered over the landscape, a little patch of snow. The sun was just up, and the whole scene was very bright and beautiful.

"Now, father," said Rollo, "tell me how you know that there is any wind at all."

"I did not say that there was any *wind*. I said *motion of the air*."

"Why, father," replied Rollo, "I thought that wind was motion of the air."

"So it is," said his father; "but all motion of the air is not wind. Wind is a *current* of air, that is, a *progressive* motion;—and in fact, there is, this morning, a slight current from the westward."

"How can you tell, father?" asked Rollo.

"By the smokes from the chimneys; don't you see that they all lean a little from the west towards the east?"

"Not but a little, father;—and there's one, from that red house, which goes up exactly straight."

"Yes," said his father, "there is one; but, in general, the columns of smoke lean; which is proof that there is a gentle current of air to the eastward."

"*Westward*, you said, father," rejoined Rollo.

"Yes, *from* the westward, but *to* the eastward.

"That is what is called a progressive motion," continued Rollo's father; "that is, the whole body of air makes progress; it advances from west to east. But there is another kind of motion, called a *vibratory* motion."

"What kind of a motion is that, father?" asked Rollo.

"It is a very hard kind to describe, at any rate," said his father. "It is a kind of quivering, which begins in one place and spreads in every direction. Don't you hear a kind of a thumping sound?"

"Yes," said Rollo, "a great way off; what is it?"

"Look over across the pond there," said his father; "don't you see that man cutting wood?"

"Yes," said Rollo; "that's what makes the noise.—No, father," he continued, after a moment's pause, "that's not it. Look, father, and you'll see that the thumping sound comes when his axe is lifted up."

They all looked, and found that it was as Rollo had said. The strokes of the axe kept time, pretty well, with the sound of blows, which they heard, only the sounds did not correspond with the descent of the axe. When the axe appeared to strike the wood, they did not hear any sound, but they did hear one every time the axe was lifted up.

"So, you see," said Rollo, "it is not that man that we hear. There must be some other man cutting wood."

"We will wait a minute," said his father, "until he gets the log cut off, and then he will stop cutting; and we will see whether we cease to hear the sound."

So they sat still, and watched the man for a minute. Presently he stopped cutting,—and, to Rollo's great surprise, the sound stopped too.

"That's strange," said Rollo.

In a moment more, the man had rolled the log over, and commenced cutting upon the other side; and in an instant after he began to cut, Rollo began to hear the sound of strokes again.

"Yes," said Rollo, "it must be his cutting that we hear; but it is very strange that he makes a noise when he lifts up his axe, and no noise when it goes down."

"I'll tell you how it is," said his father. "He makes the noise when his axe goes down; but, then, it takes some little time for the sound to get here; and by the time the sound gets here, his axe is up."

"O," said Rollo, "is that it?"

"Yes," replied his father, "that is it."

Rollo watched the motion of the axe several minutes longer in silence, and then his attention was attracted by the singing of a bird upon a tree in his father's garden, at a short distance below him.

Pretty soon, however, his mother said that it was time for her to return; and they all, accordingly, arose from their seats, and rambled along together a short distance upon the brow of the hill, but towards home.

"Then the sound moves along through the air," said Rollo, "from the man to us."

"Yes," said his father; "that is, there is a vibratory motion of the air,—a kind of quivering,—which begins where the man is, and spreads all around in every direction, until it reaches us. But there is no *progressive* motion; that is, none of the air itself, where the man is at work, leaves him, and comes to us."

"But, husband," said Rollo's mother, "I don't see how anything can come from where the man is, to us, unless it is the air itself."

"It is rather hard to understand," said his father. "But I can make an experiment with a string, when we get home, that will show you something about it."

They rambled about among the rocks for a short time longer, and then they descended by a steep and crooked path, in a different place from where they had ascended. When they had got nearly home, Rollo said that he would run forward and get his father's ball of twine and bring it out; and so have it all ready for the experiment.

Accordingly, when Rollo's father and mother arrived at the front door, they found Rollo ready there with a small ball of twine in his hand, about as large as an apple.

"Now, Rollo," said his father, "you may take hold of the end of the twine, and walk along out into the street, while I hold the ball, and let the string unwind."

Rollo did so. He drew out a long piece of twine, as long as the whole front of the house, and then he stopped to ask his father if that was enough.

"No," said his father; "walk along."

So Rollo walked on for some distance farther, until, at last, the ball was entirely unwound. Rollo had one end of it, and was standing at some distance down the road, while his father, with the other end, stood at the gate of the front yard. The middle of the string hung down pretty near to the ground.

"Draw tight, Rollo," said his father.

So Rollo pulled a little harder, and by that means drew the line straighter.

"Now," said his father, "walk along slowly."

So Rollo walked along, drawing the end of the line with him. His father followed with the other end. Thus they advanced several steps along the side of the road.

"There," said his father. "Stop. That, you see, was a *progressive* motion."

"Yes, sir," replied Rollo.

"The whole string advanced along the road," added his father. "It made progress, and so it was a progressive motion. Now, fasten your end of the string, Rollo, to that tree directly behind you."

Rollo looked behind him, and saw that he was standing near a small maple-tree, which had been planted, a few years before, by the side of the road.

"Tie it right around the stem of the tree," said his father, "about as high as your shoulder."

Rollo fastened the string as his father had directed. Then his father fastened *his* end, in the same way, to another tree, which was growing near where he was standing.

"Now," said he, "there can be no more progressive motion, but there can be a vibratory one. Take hold of the string near where it is fastened to the tree."

Rollo took hold of it, as his father had directed, and then his father told him to shut his eyes. When his eyes were shut, so that he could not see, his father said that he was going to strike the string, at his end of it, with his pencil-case, and he asked Rollo to observe whether he could feel any motion.

Rollo held very still, while his father struck the string; and immediately afterwards he called out, "Yes, sir." Then his father struck the string again, several times, and every time Rollo could feel a distinct vibratory or quivering motion, which was transmitted very rapidly through the string, from one end to the other; although, as the string was fastened by both ends to the trees, it was evident that there could be no progressive motion.

Rollo's mother had been standing all this time at the step of the door, watching the progress of the experiment; and, when she saw the expression of satisfaction upon Rollo's countenance, while he was standing, with his eyes shut, holding the end of the string, she wanted to come and take hold of it herself, so as to see what sort of a sensation the vibratory motion of the string produced.

So she came out through the gate, and asked Mr. Holiday to wait a moment while she went to where Rollo was standing, and took hold of the string. But he said that it would not be necessary for her to go there, as she could take hold of his end of the line just as well, and let Rollo strike the other end.

They accordingly performed the experiment in that way, and Rollo's mother could feel the vibrations very distinctly.

"One thing you must observe," said Mr. Holiday; "and that is, that the vibrations pass along from one end of the line to the other very quick indeed. We feel them at one end almost at the same instant that the other end is struck."

"*Exactly* at the same instant, sir," said Rollo.

"No," replied his father, "not exactly at the same instant, though it is very nearly the same."

"I did not see any difference," said Rollo.

"No," replied his father, "you cannot perceive any difference in so short a string, but if we had a string, or a wire, a mile long, I presume that we should find that it would require a sensible period of time to transmit the vibrations from one end to the other."

"What do you mean by a sensible period of time, father?" asked Rollo.

"Why, a length of time that you could perceive," said his father; "just as it was with the man cutting wood. We could see that some time elapsed between the striking of the blow, and our hearing the sound."

"Yes," said Rollo, "just as long as it took him to lift up his axe."

"That is not certain," replied his father, "because the sound that we heard might have belonged to a blow made before. That is, it might be that, when he had struck one blow, he had time to raise his axe and strike another, and then raise his axe again, before the sound of the first blow came to us."

"Yes, sir," said Rollo, "I understand."

Mr. Holiday then told Rollo that he might unfasten the string from the trees, and wind it up again into a ball, and bring it in. Then he and Rollo's mother went into the house, to see if breakfast was not almost ready.

That morning, after they were all seated at the breakfast table, Rollo said to his father that he did not exactly understand what sort of a motion the vibratory motion of the air was, after all.

"No," said his father, "I suppose you do not. And, in fact, I do not understand it very perfectly myself. I only know that the philosophers say, that, when a man strikes a blow with an axe upon a log of wood, it produces a little quivering motion of the air, which spreads all around, darting off in every direction very swiftly. If a boy strikes a tin pail with a drum-stick, it makes another kind of quivering or vibration, which is different from that which is made by the axe; but I don't know precisely how it differs. So, when the air is full of sounds, on a still morning, it is full of these little vibrations, like a string which trembles from end to end, though its ends are fastened so that it cannot move away."

"Then the air is never at rest," said Rollo's mother.

"No; certainly not, when any sound is to be heard; and it is never perfectly silent."

"There is one thing very extraordinary," said Mrs. Holiday.

"What is it?" asked Rollo's father.

"Why, that, when a great many sounds are made at the same time," she replied,—"as, for example, when we are upon the top of a hill, on a still morning, and hear a great many separate sounds, as a man cutting wood, birds singing, a bell ringing, and perhaps a man shouting to his oxen,—all those tremblings or vibrations, being in the air together, do not interfere with one another."

"Yes," said Mr. Holiday, "it is very extraordinary indeed. They do not seem to interfere at all. When there are too many sounds, or if there is a wind with them, they do interfere; but, in a calm morning, like this, when the air is at rest, you can hear a great many distant sounds very distinctly."

"Yes," said Rollo, "and I mean to go up to the top of the rocks again after breakfast, and listen."

QUESTIONS

What time of the year was it when Rollo took this walk? How did Rollo satisfy himself that there was no wind at all? How did his father prove that there was a little wind? Is all motion of the air wind? What two kinds of motion are mentioned? What sound did they hear? What made Rollo think the sound was not made by the man whom they saw cutting wood? How did his father explain this phenomenon? What experiment did they try with the string? Were the vibrations transmitted slowly or rapidly through the string? Did Rollo think that he understood perfectly the nature of the vibrations? What extraordinary circumstance did Rollo's mother mention at the breakfast table?

www.ingramcontent.com/pod-product-compliance
Lightning Source LLC
Chambersburg PA
CBHW081623100526
44590CB00021B/3581